Nano-Publics

Pat J. Gehrke

Nano-Publics

Communicating Nanotechnology Applications,
Risks, and Regulations

Pat J. Gehrke
University of South Carolina
Columbia, South Carolina, USA

ISBN 978-3-319-69610-2   ISBN 978-3-319-69611-9   (eBook)
https://doi.org/10.1007/978-3-319-69611-9

Library of Congress Control Number: 2017957087

Cover illustration: Abstract Bricks and Shadows © Stephen Bonk/Fotolia.co.uk

Printed on acid-free paper

This Palgrave Pivot imprint is published by Springer Nature
The registered company is Springer International Publishing AG
The registered company address is: Gewerbestrasse 11, 6330 Cham, Switzerland

ACKNOWLEDGMENTS

I am grateful for the support of the National Science Foundation, which provided the funding for this study.[1] A number of colleagues contributed to the maturation of this project. I am grateful to David Berube at North Carolina State University for inviting me to develop and run the public engagement events, as well as for his input on the earliest articulations of the organic public engagement methodology. I am also indebted to the work of two graduate research assistants who worked as coders on that study: Matthew Boedy and Jonathan Maricle. Finally, none of this work would have been possible without the support of my department, college, and university. The University of South Carolina has been supportive of this work, and I am especially thankful to the College of Arts and Sciences for providing additional funding that brought this project through its final months of development.

NOTE

1. National Science Foundation Grant: Nanotechnology Interdisciplinary Research Team in Intuitive Toxicology, #06-595.

CONTENTS

1 Organic Public Engagement with Nanotechnology:
   Advantages and Challenges   1

2 Public Understanding of Nanotechnology:
   How Publics Know   21

3 Nanotechnology Applications and Risks:
   Valences and Ambiguities   39

4 Government Regulation of Nanotechnology:
   Imperfectly Essential   51

5 Lessons for Science Communicators:
   Assumptions and Assessment   65

Master Works Cited   87

Index   93

LIST OF FIGURES

Fig. 2.1 Levels of public understanding   24
Fig. 2.2 Distribution of understanding by region   25
Fig. 2.3 Distribution of understanding by group type   25
Fig. 2.4 Distribution of understanding by education level   26
Fig. 2.5 Distribution of understanding by income level   26

LIST OF TABLES

Table 1.1   Most commonly occurring reliable codes   16
Table 3.1   Positive and negative audience responses to applications   42
Table 3.2   Collective disposition by application   43
Table 3.3   Weighted factors influencing disposition   43
Table 4.1   Public disposition toward government   54
Table 4.2   Public disposition toward industry   56

CHAPTER 1

Organic Public Engagement with Nanotechnology: Advantages and Challenges

Abstract Using recent advances in the study of human behavior to consider how people understand and perceive nanotechnology provides us with a richer and more actionable set of insights into public perceptions of this exciting emerging technology. At the same time, using public perceptions of nanotechnology to explore the potential of new methods of studying human behavior provides us with a case study in the power of these techniques. It demonstrates their capacity to produce insights that are not only more valid and have a greater capacity to explain human perception and action but also more useful to people who need to communicate with publics about new technologies and set regulations for those technologies.

Keywords Organic public engagement • Nanotechnology • Public understanding of science • Behavioral economics • Public engagement with science

This book is about two major technological advances: public perception of new materials engineered at the molecular level with astounding properties and the development of new techniques in the study of human behavior that are revolutionizing how we research human behavior and attitudes. The first of these goes by the broad term *nanotechnology*, while the second goes by many names but has been popularized primarily under the mantle

© The Author(s) 2018　　　　　　　　　　　　　　　　　　　　1
P.J. Gehrke, *Nano-Publics*,
https://doi.org/10.1007/978-3-319-69611-9_1

of behavioral economics. Each of these technological advances has reached the general public in just the last 20 years, with their broad adoption occurring only in the past decade. Each is on the cusp of revolutionizing industries and fields in ways that will transform them for decades to come. Each is controversial and carries its own risks along with its opportunities.

Using recent advances in the study of human behavior to consider how people understand and perceive nanotechnology provides us with a richer and more actionable set of insights into public perceptions of this exciting new technology. At the same time, using public perceptions of nanotechnology to explore the potential of new methods of studying human behavior provides us with a case study in the power of these techniques. It demonstrates their capacity to produce insights that are not only more valid and have a greater capacity to explain human perception and action but also more useful to people who need to communicate with publics about new technologies and set regulations for those technologies. This book takes on both of these tasks, offering readers an example of how to study public perception and communication of new technologies as well as a set of robust actionable insights into public perceptions and communication of nanotechnology in particular.

Nanotechnology has been a serious field of scientific and technological research for more than 30 years, but only in the past decade has it gained much public attention. In short, nanotechnology refers to "the ability to control and restructure the matter at the atomic and molecular levels in the range of approximately 1 nm to 100 nm, and exploiting the distinct properties and phenomena at that scale" (Roco 2011, p. 428). Nanotechnology derives its name from its size, measured in nanometers (nm), which are 10^{-9} meters. Conceptualizing something this tiny can be difficult, requiring comparisons to the smallest things people see and encounter on a daily basis. For example, a human hair is roughly 90,000 nm thick, and there are over 25 million nanometers in an inch. When we try to think about something 100 nm in size, we are trying to imagine something much smaller than most of us have ever seen, even through a microscope. Germs are typically about 1000 nm, and red blood cells tend to be about 8000 nm in diameter. So, if something must be between 1 nm and 100 nm to qualify as a nanotechnology, it has to be no larger than one-tenth the size of a germ and one-eightieth the size of a red blood cell.

As fantastic as it may sound to think that we could engineer technologies at such a tiny scale, most of us use technologies manufactured at the nanoscale every day. Modern computers and smartphones include processors that were manufactured with processes that boast levels of precision ranging from 10 nm to 32 nm, and have been manufactured at the 90 nm or smaller scale since 2004. However, some researchers do not consider these processors true examples of nanotechnology. While there are advantages to making processors with these finer levels of precision, nanotechnology researchers argue that they do not "exploit the distinct properties and phenomena" that occur when you make particles so incredibly small.

Instead, Michael Roco (2011) and others prefer to reserve the term *nanotechnology* for materials like carbon nanotubes, which can be built with a wide variety of unique and sometimes astounding capacities depending on their shape and form. As the name implies, these are nanoscale tubes made of carbon atoms. They can be made incredibly strong and stiff, perhaps being the strongest material yet invented while still being quite light. How well they conduct electricity and heat can be manipulated by using a different pattern of carbon atoms to form the walls of the tube, some promising conductivity hundreds of times greater than copper. These kinds of properties are possible only because of manipulation of matter at the atomic and molecular level that exploits the possibilities that emerge at such a small scale.

In the past decade, these materials have gradually worked their way into almost every segment of consumer products. Sports equipment like golf clubs and baseball bats now sometimes include carbon nanofibers to make them stronger and more durable. Some paints include nanomaterials made from titanium dioxide to reduce how much heat they absorb, make them easier to clean, or less flammable. A wide variety of materials are manufactured in nanoscale and incorporated into products, including silver nanomaterials, because, at the nanoscale, silver has antibacterial properties. Nanoscale titanium dioxide has also been a very popular addition to sunscreens and cosmetics, because it not only provides good protection against sun exposure but also, at the nanoscale, tends to refract light like a prism, giving the effect of a glow or sheen. In recent years, some food packaging has been manufactured with nanomaterials that slow bacterial growth or make the packaging stronger.

Yet, for all its potential and even with the widespread adoption of nanomaterials in consumer products, public awareness remains quite limited. Most people display little knowledge about nanotechnology and are unaware that the products they use contain nanomaterials. The government has sponsored numerous public events and information campaigns about the technology, and researchers have been exploring public perceptions for many years. All the same, even for all its promise and its ubiquity in our daily lives, most people are neither excited about nanotechnology nor concerned about their exposure to it.

Most of the data we have about public understanding and perception of nanotechnology comes from survey research, some of which was conducted at or following public engagement events on the topic. While that data has been helpful in drawing a baseline picture of diverse publics and how they have (or have not) encountered and responded to these emerging technologies, we have very little research that reflects the paradigm revolution happening in the methods of studying human behavior.

Thanks to popular books by academics like Daniel Kahneman (2011) and Steven Levitt (Levitt and Dubner 2005), behavioral economics has been featured in popular magazines, national and local newspapers, radio and television shows, and countless blogs and podcasts. While much of that work comes from economists (like Levitt), psychologists like Kahneman have also been key to revolutionizing economic theory. In fact, Kahneman is the first PhD in psychology to ever win the Nobel Prize in economics, which he received in 2002 for his work with Amos Tversky on behavioral economics.

What Kahneman, Tversky, Levitt, and others in this revolution share is a focus on the empirical study of human behavior as it actually occurs (rather than as it occurs in a laboratory setting or should occur in an assumed world of rational choice, perfect information, or similar counterfactual conditions). That focus is certainly not new, and certain branches of psychology and other social sciences have sought such a turn since almost the beginning of the social sciences themselves. As Adam Lerner and I documented in our recent book (Lerner and Gehrke 2018), the origins of the current revolution in the study of human behavior can be traced back to the human ecology movement and the Chicago school of sociology in the 1920s. Throughout the twentieth century, various strains of ecological thinking have resisted social science's previously dominant paradigm of relying upon laboratory experiments, tightly controlled environments, and survey data. Yet, the true paradigm shift in the study of human behavior did not begin until those same tightly controlled laboratory experiments demonstrated their own insufficiency. Since the 1990s, and especially in the most recent two decades, researchers in cognitive psychology, economics, education, jurisprudence, political science, and nearly every other social scientific discipline have demonstrated that humans are not predominantly rational actors, often affected not only by their own emotions and biographies but also by elements of their environment ranging from the most overt influences to the most subtle cues. Efforts to explain away these findings with theories of rational actors, logical choice, deliberative principles, or complex motivations do little more than build precarious paper houses that topple under the lightest breeze of scrutiny.

However, for all their revelations about the myriad nonrational elements in human behavior and even with their frequent findings that people's choices and actions are strongly influenced by factors like scene and setting, or medium of communication, or perceived outcomes, behavioral economists still sometimes deploy laboratory research methods where the scene, setting, medium of communication, or perceived outcomes are contrived and deviate from the world in which research subjects actually act.

The irony is, and this is true in a wide variety of social sciences, that their research outcomes frequently indict their research methods. If it is true that the configuration and comfort of seating affect restaurant diners (Robson 2002), then any researcher studying dining behavior would need to ensure the chairs and configuration match the places where people actually dine. If time of day affects how we judge of other people, such as in parole hearings (Danzinger et al. 2011), then researchers would need to account for this in all cases where they study judgment. In short, as myriad cues and influences come to light and are documented by behavioral economics and cognitive psychology, the more variables a researcher must account for to produce valid research.

This problem is not a reason to condemn the social sciences or to demean behavioral economics and cognitive psychology. Quite the opposite, they demonstrate how much these fields have done to prove the power of our environment, the whole ecology we live in, to influence our behavior. The true paradigm shift in the study of human behavior is not so much between the rational and behavioral (be it in economics or another field) but between those who study behaviors in contrived artificial environments and those who study behaviors as they occur in ecologies outside the laboratory.

The current paradigm shift afoot is a recognition that the best way to account for the countless diverse aspects of an environment that affect human behavior is to study that behavior *in situ*, in the place and situation where it occurs outside of the study design. When this is not possible, researchers ought to give high priority to strengthening the ecological validity of the research design. *Ecological validity* can be defined as "the degree of correspondence between the research conditions and the phenomenon being studied as it occurs naturally or outside of the research setting" (Gehrke 2018). Ecological validity has been a fundamental criterion for good science since the middle of the twentieth century. Likewise, the principles behind ecological validity have driven some of the rise of empirical user experience research over the previous dominance of user interface design (e.g., Roto 2015). Yet, under an old-paradigm's quantitative experimental models of social science, ecological validity is too often mistakenly folded into the principle of external validity.

In practice, demonstrating sufficient external validity for publication today often does not require any consideration of ecological validity or any correspondence to the actual conditions under which a studied phenomenon occurs outside contrived research settings. Today researchers argue for their external validity simply by claiming their sample is sufficiently representative of the population to which they generalize. Or they presume external validity absent the occurrence of a dramatic event or condition that makes the setting of the study different from the normal place and time in which the broader population lives. In some cases, researchers will even claim external validity because their study matches the artificial laboratory conditions set in other researchers' studies. None of these attempts to validate external validity account for the challenge posed by the last few decades' research on the effects of environment on human decision and action.

The error of folding ecological validity into external validity has allowed this slippage, wherein researchers are rigorous in their sampling methods but do little more than wave their hands in the direction of the effect that environment can have on their subjects. In psychology, this has created a rift between the clinical psychologists, who tend to work closely with a small number of patients living through the actual processes being studied, and research psychologists, who tend to study large numbers of people but most commonly do so in settings divorced from the lived conditions of the subjects. The clinical psychologists are criticized for poor sampling and the research psychologists for lacking ecological validity, leaving both arguing the other has generated unreliable results.

Yet, for all the sparks generated by this debate, few seem to recognize that the work of the research psychologists keeps demonstrating the truth of the clinical psychologists' critique. In study after study, nearly every aspect of an environment has been shown to impact various behaviors:

how you pose a question (Lave 1997), the color of the room (Alter 2014), the lighting (van Bommel and van den Beld 2004), medium of communication (Bornstein 1999), and the list could go on for dozens of pages. How is a laboratory researcher possibly going to account for all of those challenges to their ecological validity? The answer is that they cannot. In fact, it would be far easier to manage the sampling problem faced by field researchers (such as clinical psychologists), which is exactly the direction that current social sciences are headed.

We are still in the midst of this paradigm shift in the social sciences, as we can see from a perusal of recent publications in academic journals. Most still deploy the kinds of analytic, experimental, and contrived methodologies that were pioneered in the 1950s and 1960s, advanced in recent decades largely by the arrival of computers and statistical methods capable of handling larger data sets. Even in behavioral economics and behavioral psychology, one finds published studies that seem to have forgotten that the environment in which the research takes place can dramatically alter the behavior of participants and subjects.

This struggle to move past the research methods of the last century and go beyond their intrinsic failure to generate strong ecological validity is well demonstrated by current debates in public engagement, especially public engagement with science. In 2007, I began development of a methodology for researching public engagement with emerging technologies that accounted not only for the findings on the effect of environment and setting, but was grounded in theories more attentive to the full communication ecology in which public engagement occurs. A year later, I was invited to Washington, D.C., to attend a meeting of National Science Foundation (NSF) primary investigators and present my argument for that methodology. In a room of about 50 scholars and researchers from around the country and a number of NSF officials, I laid out my case for the importance of ecological validity in public engagement research and the drawbacks of many of the old-paradigm approaches that were still too commonly used. I argued for a new model of public engagement, which I have called *organic public engagement* (Gehrke 2014; Lerner and Gehrke 2018). The reception was not simply mixed, but was polarized. Roughly half the room was excited and applauded the innovations and paradigm shift I mapped out during my talk. The other half behaved as if I had threatened to defile their holy temples and burn their sacred texts. To be fair, when one presents a methodological argument in broad terms during early stages of development, it can often be received as sacrilegious and arrogant, and my presentation then did little to mitigate that kind of reaction. More importantly, I saw that the most ardent and vocal opponents of organic public engagement were also researchers and scholars who had a significant financial interest in receiving or retaining NSF grants and whose training and background had focused heavily on old-paradigm methodologies of studying human behavior.

Over the next three years, I deployed, tested, and refined the organic public engagement methodology in 11 different public groups all across the United States. As a coprimary investigator on a nanotechnology interdisciplinary research team, funded by the NSF, my job was to design, carry out, and analyze public engagement events that would help us better understand public perception of the risks of nanomaterials and nanotechnologies. That meant deciding whom to study, where, by what method, and in what kind of an engagement format. My colleagues thankfully trusted me enough to give me a free hand in the process, and as a result, it became the first test case for the organic public engagement methodology.

The core principles of organic public engagement and the argument for it can be found both in my 2014 essay in *Public Understanding of Science* and in a recent book I coauthored with my doctoral advisee, Adam Lerner (Lerner and Gehrke 2018). Unlike old-paradigm methods that understand sampling as picking out individuals from a population, organic public engagement tends to sample by locating publics and public ecologies where engagement with the topic in question is already happening or might occur naturally in the future. This kind of deliberate sampling not only allows us to focus research on ecologies and populations where engagement occurs outside the research ("in the wild") but also helps locate potential spaces that will serve to shape and lead opinion on the topic. In the public engagement with nanotechnology events that I conducted, we began by asking where in the country we might find a heightened interest in nanotechnology or where such an interest might develop. That is a tricky question and not best handled by speculation, but fortunately, Philip Shapira and Jan Youtie (2008) at The Georgia Institute of Technology had developed a set of maps that located where work and innovations in nanotechnologies were clustered in the United States. Additional mapping projects at the University of Arizona (Li et al. 2009) and the Woodrow Wilson Center (nanotechproject.org) provide additional details and updated information. Most such maps also provided a classification of different kinds of nanotechnology development, such as government, university, and private industry. I decided we should try to get some representation from each of these kinds of development nodes, as a single geographic area often contained multiple types of nanotechnology development, and such areas were prioritized in locating engagement events.

The maps of nanotechnology development showed a wide variety of locations across the country, so I wanted to reflect that diversity in how I selected sites for this study as well. I divided the map following U.S. Census Bureau-designated regions: West, Midwest, Northeast, and South. Balancing this regional difference with the maps of nanotechnology development meant not every region would receive equal representation in the study, but each region needed at least one event, and ideally two. Additionally, since my home base is in the South region, it was easier and less expensive to access sites in the South. Scarcity of resources is an unfortunate but frequent limitation in research methods, and this study was no exception.

I then made a list of the most common current and forthcoming applications of nanotechnology and nanomaterials. Locating these into clusters such as consumer discretionary goods, automotive applications, energy and efficiency, electronics, and medical applications generated a picture of the kinds of public groups, associations, and communities that might be interested in an engagement event with nanotechnology. I then conducted an Internet search using Google for sites and news stories about nanotechnology, nanomaterials, or nanoparticles that featured public interest or community groups. From this data, I decided to focus my sampling toward five different broad types of public groups: technology clubs or associations, civic and philanthropic societies with interest in areas where nanotechnology held promise, consumer interest or advocacy groups, and continuing education groups in fields or areas where nanotechnology was entering use.

At this stage, my team and I began reaching out to groups that met theses sampling criteria. Unlike old-paradigm social scientific sampling, this meant we were focused on sampling ecologies, not individuals. We had selected those ecologies deliberately because they were representative of where nanotechnology either was already in or was likely to enter public awareness. While the most common form of social science sampling would have been to recruit individual participants and then gather them together for an engagement event or similar form of research, in this study, we intentionally chose to sample whole existing associations and groups.

After all, if we know that something as simple as the color of walls or the temperature of the room may affect human behavior and attitudes, then certainly altering the social and communicative aspects of an ecology could be disastrous for the ecological validity of the research.

Plucking someone out of their normal social groups and asking them to communicate with strangers, likely in a social dynamic dramatically different from their daily communication, and then expect them to form opinions and behave in ways that somehow correspond to what they do outside that contrived setting is, at best, naïve. Adam Lerner and I have extensively documented this problem with most public engagement events (Lerner and Gehrke 2018), and it seems intuitively obvious to the casual observer, so I won't belabor the issue here. One will have no difficulty in finding hundreds of studies over decades of work that evidence that people behave differently when they encounter different people or encounter each other under different social or communicative settings (e.g., Locher et al. 2005; Phillips et al. 2006; Sundstrom 1975; Zhang 2008).

One element of sampling shared by this study and the vast majority of old-paradigm social scientific work is the voluntary nature of participation and resulting effects of self-selection. Participants are rarely tricked or forced into participation in research; only people willing to volunteer for the study after being informed of its scope and purpose are included in the sample. These are basic ethical standards in all human-subject research and only very rarely will researchers be allowed to deviate from these requirements. As a result, research participant samples almost always are not representative of the broader population, often reflecting a subset of the population with a higher interest in the topic, a greater susceptibility to the research incentive, or simply more free time. Because old-paradigm social science relies on random and broad sampling to argue for its external validity, this self-selection bias is cast as a problem that must somehow be mitigated. This is just one example of a long-standing tension between standards of ethical research and old-paradigm social science.

However, when one moves from random sampling with a focus on sampling validity to deliberate sampling with a focus on ecological validity, the problem of self-selection bias can disappear. In fact, new-paradigm social science in general and organic public engagement in particular can leverage self-selection bias to improve the quality of the sample. For example, in this study we were especially interested in studying public groups where nanotechnology already was or was likely to become a topic of interest. When we contacted groups and associations (by email or telephone, depending on their preferred mode of communication), we extended a short, scripted, simple invitation. We informed them we were studying how different public groups might talk about nanotechnology and we believed their group might have an interest in the topic. We did not provide any definition of the term and did not attempt to persuade them they should be interested. Some of the groups we contacted were highly organized and, often, local branches of national associations with stable membership. Others were loosely affiliated groups of people who had a shared interest and recurring meetings or events but no formal membership. Some were associations that organized public meetings and events with frequently recurring participants but neither membership nor regularly recurring meetings.

We offered a modest incentive of $500 to the group if it would dedicate part of an existing, planned, regular meeting of the group to the topic or organize a special event on the topic for members or regular participants. We had a number of discussions about how much this incentive might draw in public groups with no real interest in the topic, reducing the positive value of self-selection bias in our deliberate sampling method. We certainly did not want groups who had no interest and were unlikely to develop any interest in the topic signing on to the study just for the money. In the end, we made three choices that mitigated this risk. First, when we offered the incentive, we framed it as funds to offset the costs of organizing and hosting the event. Second, we expressed clearly that we expected them to handle all the planning, logistics, and communication related to the event. The research team would not be directly involved in the event planning or implementation. We also considered this important to preserve the ecological validity of the research. Third, we limited the incentive payment to $500, which is a very modest sum even for a small local civic organization, especially relative to the work we were asking them to do. Over two-thirds of the groups we contacted declined to participate.

Eleven groups accepted our invitation to participate. Three were philanthropic or service societies, two were consumer advocacy or interest groups, two were continuing education groups, two were college student associations, one was a technology and gaming convention, and one was a religious group. Five were located in rural areas or small towns, four were in large cities, and three were in mid-size cities. Five were located in the South, three in the Midwest, two in the West, and one in the Northeast. We found a number of factors appeared to affect groups' willingness to participate that future organic public engagement research may want to mitigate or account for.

First, geographic proximity to the person extending the invitation or institutions affiliated with the study appeared to affect likelihood to accept the invitation. One of the reasons the South is overrepresented in this study may be because I am affiliated with the University of South Carolina and the other primary institutional home for the grant was North Carolina State University. Second, personal connection or affiliation between someone on the research team and a public group dramatically increased the likelihood of a group accepting our invitation. While none of the researchers involved with this study were members of or directly participated in any of the groups who accepted our invitation, in two cases, a colleague, friend, or family member of a researcher had some connection to that group and facilitated the invitation. I debated the merits of inviting such groups for some time, but decided to use them for two reasons. First, of course, is expediency, which is often a major factor in social scientific research (even if few researchers will openly admit such). However, expediency alone would not have justified using such groups if they would reduce the validity of the research. In the end, I was persuaded by a colleague's argument that if, in fact, we were primarily interested in studying groups who were already interested in nanotechnology or who might be likely to develop such an interest, then groups with some connection to a researcher working in that field seem a logical category of our target population. In the end, that argument not only persuaded me to include these two groups but was also vindicated by finding that two additional groups who had no known connection to any of the members of the research team and certainly no connection to the person who extended the invitation still turned out to have a close connection to someone who was doing work in nanotechnology.

We encouraged each cooperating association or organization to build its own event or meeting on an issue in nanotechnology or the nanosciences that might interest its members and relate to their shared goals or purposes. In most cases, the cooperating association or organization developed its own event, located outside speakers (if it chose to use such), and handled all of the logistics. There was often some early-stage dependency on information from the research team about nanotechnology, as contacts and members of the organization or association had little familiarity with the topic. In that stage, we sought to understand the interests and goals of the organization first and then suggested at least five applications or dimensions of nanotechnology that related to the goals or interests reported by the group.

In three cases, we were asked to assist in finding speakers. We began by reviewing recent previous speakers at the cooperating association or organization and asking about its local norms and preferences. We then offered at least three broad categories of speakers and used the information provided by the local association as a guide. Once a speaker was found who fit the cooperating organization's criteria, we put the organization in contact with the speaker and stepped out of the conversation. Again, in most cases, each cooperating organization or association chose its own topic, found its own speaker, and organized its own event. In all but one event, speakers external to the cooperating association or organization were brought in as "guests" or "experts." In two cases, panels of multiple speakers were assembled; in eight cases, a single speaker was brought in; and in one case, the members of the organization educated themselves and held their own informational meeting and discussion.

When an organization chose an outside speaker, prior connections (even if tangential) were strong factors in the selection process. Local geographic presence and affiliation with well-known local research facilities and universities were also strong factors in speaker selection. Overall, of the ten events that featured outside speakers, eight used university researchers, one used researchers from a national laboratory, and one used both university and national laboratory researchers. Most of the featured researchers were primarily identified (and self-identified) as scientists, holding faculty positions in fields such as chemistry, physics, biology, and the like. However, a small minority of faculty featured in these events held degrees in anthropology, philosophy, or political science.

Approximately two weeks before a typical event, when the cooperating organization or association was comfortable with the procedure, we distributed surveys to the members, on paper with return envelopes, by email as web surveys, or in both formats, depending on the group's preference. Eight of the 11 organizations allowed and facilitated these pre-event surveys, while three either asked us not to distribute them or could not facilitate their distribution. We counted only surveys completed before an event in our data.

The dominant political views of the public groups ranged from liberal to conservative, including groups that self-identified as moderate. Member age in these groups ranged from 20 to 74, while gender composition varied from almost evenly split to strongly male dominated to strongly female dominated. Average education level and family income were skewed to the higher end of the spectrum. The lowest average education in any group was some college and the highest average was some graduate education (all participants reported at least a high school diploma or equivalent). The group with the lowest average family income reported 50–75 thousand dollars per year, while the group with the highest average reported between 75 and 100 thousand and 100–150 thousand. However, within some groups, we found great variety in income and education levels, with some representing both the highest and lowest income levels. We did not collect data on race or ethnicity.

The public groups structured their events in relatively traditional ways. In most cases, a speaker or a panel of speakers made a prepared presentation, lasting between 20 and 50 minutes, depending on the norms of the groups. Likewise, most had question-and-answer periods or open discussion following the presentations. With two exceptions, the speakers used slides, to provide graphic and textual support to their spoken presentations, as was the practice for featured speakers under normal circumstances. One of the core research team members observed and recorded all events except one, which was recorded by a proxy who took extensive field notes for the researchers. We recorded ten events using audio and video and one using audio only because of confidentiality concerns among that group's members.

We then converted each recording to a standard format, overlaying the video with a confidentiality statement, and analyzed the results using computer-assisted qualitative analysis. The process of computer-assisted qualitative analysis allowed four researchers to view or listen to all 11 events and develop a list of the most prominent and important topics, behaviors, and attitudes in the recordings. We developed 29 general categories, such as attitudes toward government, human health impacts, metaphors and similes, moral and ethical appeals, speaker expertise, and the use of visual aids. We then subdivided each category into more specific elements. For example, we categorized attitude toward government first into positive, neutral, and negative and then divided further based on the reasons behind the attitudes (such as negative due to questions of competency, goodwill, slowness, and unfunded mandates). These became the codes we assigned to moments in the event recordings during the next phase of analysis.

We then divided the video and audio recordings into units that reflected topical turns and moments in the larger events. Most units of analysis were roughly a minute long. We coded units based on who was speaking (presenter or audience); next, two members of the team, working independently, coded each unit using the codes previously established.

We used a Computer-Assisted Qualitative Data Analysis tool called NVivo[1] to facilitate coding and to allow us to code directly on the video (rather than requiring transcripts). That software also allowed us to compare the reliability between these coders as a basis for confidence in our findings. To assess reliability, we used Cohen's Kappa Coefficient, which compares the actual agreement between the coders to the random chance of agreement, and assigns a score to each code based on the result. Table 1.1 presents the resulting list of the most commonly occurring reliable codes.

Based upon the coding found to be most reliable, we extracted video and audio of the most reliable and prominent codes and then studied those moments both for relationship to other codes and for their broader characteristics. From this analysis, we isolated the most common topics and themes, which we defined as occurring in at least seven of the 11 events and at least 20 times total across the events, and for which our coding was significantly reliable. We retained for comparison the codes that were at least moderately reliable but less frequent, and we removed from further analysis the codes that were less than moderately reliable.

This allowed us to find which codes commonly occurred with other codes but also which codes demonstrated complex associations, such as applications for cancer treatment having both positive and negative audience responses. Filtering our codes to only those with good or very good reliability and with the outputs on commonly associated codes from NVivo, I was then able to return to key moments in each video that shared the same audience reaction, topic, or other code element and interpret what the code relationships could tell us about public perception of nanotechnology.

The following chapters provide results and discussion. Chapter 2 explores what and how publics know about nanotechnology. While most research on public understanding of nanotechnology reports little to no awareness, such research has focused on individuals rather than on the collective knowledge of community groups and civic associations. Given that our voluntary associations in civil society form important networks of information and incubators of opinion, collective knowledge within community groups may be more significant in determining public response than that of individuals. Through a combination of open-ended survey questions and observation of public engagement events, this study explored the presence and operation of such collective knowledge of nanotechnology. We found that publics display not only a high potential for good foundational knowledge, but also a capacity to rapidly build knowledge competencies when motivated.

Table 1.1   Most commonly occurring reliable codes

Code	No. of events	Occurrences	Reliability
Nanotech is like asbestos	7	25	0.89 (very good)
Nanotech is like human hair	8	24	0.95 (very good)
Applications: antimicrobial	7	20	1.0 (very good)
Applications: computers & smartphones	9	60	0.92 (very good)
Applications: environmental remediation	7	21	0.98 (very good)
Applications: cancer treatment	7	34	0.92 (very good)
Applications: drugs & drug delivery	10	70	0.89 (very good)
Applications: medical imaging & sensing	10	34	0.86 (very good)
Applications: energy efficiency	8	39	0.99 (very good)
Applications: sensors (generic or undefined)	9	34	0.98 (very good)
Applications: solar power	7	28	0.89 (very good)
Applications: sports equipment	7	23	0.84 (very good)
Applications: sunscreens	7	58	0.84 (very good)
Applications: textiles & fabrics	9	71	0.92 (very good)
Applications: water purification or desalination	7	30	0.81 (very good)
Attitudes toward government: negative	8	91	0.75 (good)
Attitudes toward government: positive	8	45	0.75 (good)
Attitudes toward general technology: cautious	10	161	0.88 (very good)
Attitudes toward general technology: optimistic	11	142	0.87 (very good)
Audience reaction: laughter	10	245	0.90 (very good)
Audience reaction: vocalized assent	8	31	0.79 (good)

(*continued*)

Table 1.1 (continued)

Code	No. of events	Occurrences	Reliability
Definitions of nanotech: active vs. passive	7	30	1.0 (very good)
Definitions of nanotech: control surface area	8	43	0.82 (very good)
Definitions of nanotech: invisible to human eye	8	28	1.0 (very good)
Definitions of nanotech: new & revolutionary	9	40	0.82 (very good)
Definitions of nanotech: not new	9	27	0.89 (very good)
Definitions of nanotech scalar (small, 10^{-9})	11	80	0.92 (very good)
Market size for nanotech	7	43	0.94 (very good)
Moving to market vs risk assessment	9	41	0.98 (very good)
Harms to environment (generic, undefined)	7	57	0.96 (very good)
Nanotech is "futuristic" or "science fiction"	7	44	0.81 (very good)
Human health risks of nanotech (unspecified)	7	52	0.95 (very good)
Nanotech intentionally made by humans	9	52	0.96 (very good)
Nanotech made by nature	10	39	0.97 (very good)
Nanotech made from carbon	8	56	0.98 (very good)
Nanotech made from gold	7	41	0.97 (very good)
Moral or ethical appeals	10	126	0.83 (very good)
Nanotech is needed for efficiency	9	36	0.80 (very good)
Nanotech is needed for progress (in general)	7	23	0.80 (very good)
Pathetic/emotional appeals	8	59	0.98 (very good)
Shapes of nanotech: tubes	7	45	0.88 (very good)

(*continued*)

Table 1.1 (continued)

Code	No. of events	Occurrences	Reliability
Speaker shares a cause or goal with audience	9	20	0.74 (good)
Speaker shares geography or locale with audience	10	41	0.85 (very good)
Speaker speaks as fellow citizen or lay person	8	60	0.72 (good)
Speaker acknowledges limits of own knowledge	8	51	0.89 (good)
Speaker answers a question especially well	8	44	0.76 (good)

Chapter 3 examines particular applications of nanotechnology and associated perceptions of risk. The reliable codes include 13 applications of nanomaterials that were most commonly mentioned at the 11 events. We assessed audience concern by examining conversations that included both mention of one or more of those 13 applications and some form of positive or negative response from the audience. Within the scope of audience response, we included verbal statements, the tone or disposition of questions, and nonverbal or paralinguistic responses (such as audience nodding in agreement or making audible sounds of assent). Our analysis of expressions of concern or interest (both positive and negative) among audience members indicated that nine of the 13 applications were of the highest concern for these public groups. We then took the previously assessed positive and negative disposition codes and compared them to each instance of audience response. This allowed us to produce a positive and negative disposition score for each application and to break out specific subtopics of audience concern.

Chapter 4 examines the public groups' expressed sentiments about regulation of nanotechnology. Overall, publics' responses to nanomaterials and nanoparticles in consumer products were more favorable to government regulation than not. As more than one attendee put it, consumers cannot see the nanomaterials and have no way to test for them, no labeling informs consumers of their use, and in many cases, there is little to no regulation or required testing. Add to this the near consensus among audiences and experts alike that testing and research on the risks of exposure (particularly long-term effects) are inconclusive, insufficient, or at best conflicted, and one would have to place an enormous amount of faith in private actors (particularly large corporations) to conclude that such regulation and oversight does not need strengthening. In the case of nanomaterials in cosmetics and sunscreens, our research found conditions favorable to regulation to a degree not present in any other product category for the following reasons: a lack of positive perceptions (benefits), the near impossibility of consumer choice affecting exposure (due to lack of labeling), high uncertainty about the health and safety risks, and the likelihood of regular and persistent exposure to the products. At the same time, audience members expressed skepticism of the competence and benevolence of government actors and regulatory bodies. Analysis of specific conversations provides nuance to this tension.

Chapter 5 concludes the book with lessons for science communicators and engagement event organizers. Combing the results from Chaps. 2, 3 and 4 with analysis of specific behaviors of experts at these events and audience reactions produced guidelines for science communicators. Perhaps most significant is the problem of science experts' poor understanding of the public groups they spoke with. In many cases, experts underestimated a public group's knowledge and level of engagement, in ways that were seriously damaging in a few cases to their communication objectives. Experts also sometimes demonstrated poor understanding of the genre or format of the communication event and were exceedingly academic and dry in both content and delivery, including visual aids. These two areas pose special problems for public communication of science and technology, and I offer advice for future expert speakers.

NOTE

1. Neither the author nor any member of the research team has an affiliation with QSR International, the publisher of NVivo.

WORKS CITED

Alter, A. (2014). *Drunk tank pink: And other unexpected forces that shape how we think, feel, and behave.* New York: Penguin Books.

Bornstein, B. H. (1999). The ecological validity of jury simulations: Is the jury still out? *Law and Human Behavior, 23*(1), 75–91.

Danzinger, S., Levav, J., & Avnaim-Pesso, L. (2011). Extraneous factors in judicial decisions. *Proceedings of the National Academy of Science, 108*(17), 6889–6892.

Gehrke, P. J. (2014). Ecological validity and the study of publics: The case for organic public engagement methods. *Public understanding of science, 23*(1), 77–91.

Gehrke, P. J. (2018). Ecological validity. In B. Frey (Ed.), *The SAGE encyclopedia of educational research, measurement, and evaluation.* Thousand Oaks: Sage.

Kahneman, D. (2011). *Thinking fast and slow*. New York: Farrar, Straus, & Giroux.

Lave, J. (1997). What's special about experiments as contexts for thinking. In M. Cole, Y. Engstrom, & O. Vasquez (Eds.), *Mind, culture, and activity: Seminal papers from the Laboratory of Comparative Human Cognition* (pp. 57–69). Cambridge: Cambridge University Press.

Lerner, A. S., & Gehrke, P. J. (2018). *Organic public engagement: How ecological thinking transforms public engagement with science*. New York: Palgrave Macmillan.

Levitt, S. D., & Dubner, S. J. (2005). *Freakonomics: A rogue economist explores the hidden side of everything*. New York: Harper Collins.

Li, X., Hu, D., Dang, Y., Chen, H., Larson, C. A., & Chan, J. (2009). Nano Mapper: An Internet knowledge mapping system for nanotechnology development. *Journal of Nanoparticle Research, 11*(3), 529–552.

Locher, J. L., Robinson, C. O., Roth, D. L., Ritchie, C. S., & Burgio, K. L. (2005). The effect of the presence of others on caloric intake in homebound older adults. *Journal of Gerontology, 60*(11), 1475–1478.

Phillips, A. C., Carroll, D., Hunt, K., & Der, G. (2006). The effects of the spontaneous presence of a spouse/partner and others on cardiovascular reactions to an acute psychological challenge. *Psychophysiology, 43*(6), 633–640.

Robson, K. A. (2002). A review of psychological and cultural effects on seating behavior and their application to foodservice settings. *Journal of Foodservice Business Research, 5*(2), 89–107.

Roco, M. C. (2011). The long view of nanotechnology development: The National Nanotechnology Initiative at 10 years. *Journal of Nanoparticle Research, 13*(2), 427–445.

Roto, V. (2015). *Ecological UX studies*. Paper presented at CHI'15 workshop ecological perspectives in HCI: Promise, problems, and potential. Seoul. Retrieved from http://rizzo.media.unisi.it/EPCHI2015/resources/papers/EcologicalUXstudies.pdf

Shapira, P., & Youtie, J. (2008). Emergence of nanodistricts in the United States: Path dependency or new opportunities? *Economic Development Quarterly, 22*(3), 187–199.

Sundstrom, E. (1975). An experimental study of crowding: Effects of room size, intrusion, and goal blocking on nonverbal behavior, self-disclosure, and self-reported stress. *Journal of Personality and Social Psychology, 32*(4), 645–654.

van Bommel, W. J. M., & van den Beld, G. J. (2004). Lighting for work: A review of visual and biological effects. *Lighting Research & Technology, 36*(4), 255–266.

Zhang, Y. (2008). The effects of perceived fairness and communication on honesty and collusion in a multi-agent setting. *The Accounting Review, 83*(4), 1125–1146.

CHAPTER 2

Public Understanding of Nanotechnology: How Publics Know

Abstract Most previous studies of public understanding of nanotechnology have found that very few members of the public know much about this emerging field. Such results set the expectation among scientists, regulators, industry, and science communicators that publics are largely ignorant of nanotechnology and, hence, unprepared to engage in a discussion about research priorities or regulation. However, things may not be as grim as these studies and the engagement skeptics make out. Because these studies have largely relied on survey research to determine public understanding, they have captured data that reflects only what individuals know in isolation (the condition in which they tend to answer such surveys). For precisely this reason, these studies tend to miss not only the "wisdom of the crowd" but even the collective knowledge and cooperative ways to reasoning that make up most of how people think and live their lives.

Keywords Nanotechnology • Public understanding of science • Wisdom of crowds • Information deficit

Most previous studies of public understanding of nanotechnology have found that very few members of the public know much about this emerging field. When researchers asked people to self-report their level of knowledge about nanotechnology, they found varied but generally poor public understanding. Scheufele and Lewenstein (2005) report that only 16% of respondents felt at least "somewhat informed" about nanotechnology (p. 662). Cobb and Macoubrie (2004) report that 80% of their respondents said they had heard little or nothing (pp. 299–301). A study published by Macoubrie (2006) two years later indicated even lower levels of knowledge, with 56% reporting they had heard nothing and 39% reporting only a little. Hart Research Associates (2007) reports that the large majority of Americans have heard little (27%) or nothing (42%) (p. 6). Even among demographics most likely to be using products containing nanoparticles, they found very little knowledge (p. 1).

Such results set the expectation among scientists, regulators, industry, and science communicators that publics are largely ignorant of nanotechnology and, hence, unprepared to engage in a discussion about research priorities or regulation. As Miller (2008) put it, such very low levels of public awareness mean that "we have limited information regarding which aspects of nanotechnology concern members of the public, and what action they would like to see taken in response" (p. 275). Skeptics of public engagement with science sometimes point to low public knowledge to argue that the "deficit in public understanding" limits the ability of public engagement to tap into mass opinion or the "wisdom of the crowd" (e.g., Johnson 2009).

However, things may not be as grim as these studies and the engagement skeptics make out. Because these studies have largely relied on survey research to determine public understanding, they have captured data that reflects only what individuals know in isolation (the condition in which they tend to answer such surveys). For precisely this reason, these studies tend to miss not only the "wisdom of the crowd" but even the collective knowledge and cooperative ways to reasoning that make up most of how people think and live their lives. Even in a small group of 15 people, the dynamic capacities of the group dramatically exceed that of not only any one individual, but even of the aggregate knowledge and abilities of those individuals. Groups often demonstrate synergies (both positive and negative) that allow them to display much higher levels of understanding, greater learning capacity, and more resources for reasoning than isolated individuals.

Most research on public understanding of science misses this somewhat obvious fact because it works at the level of individual knowledge and then extrapolates to assign those qualities to a public by simple aggregation. This is somewhat like trying to understand a human body by studying the qualities of its parts and then just adding up the most frequently occurring bits and assigning those qualities to a human being. In some regard, this may be accurate, but it misses everything that matters, the ways the parts all interact, support each other, provide balance, and serve different functions. The only way to study such elements is to watch the body as a whole and the interactions of the parts as the body moves and functions. This is no less true of studying people as social, communicative, and political beings, that is to say, studying publics. Only by observing the operation of a public group and the interaction of its constituents can we say to have learned much about what those public groups know and how well they can learn, discover, and reason about any topic, including nanotechnology.

Thus, in designing this study, we sought to explore the distance between aggregated survey data and the kinds of knowledge actually displayed by the publics in their engagement events. Roughly two weeks before each group's planned event on nanotechnology, we surveyed its members and likely event attendees. The first question on this survey was an open-ended prompt: "What comes to mind when you hear the word 'nanotechnology'?" The terms *nano* and *nanotechnology* were not defined or given framing anywhere in the remainder of the survey, which focused more generally on science and technology as well as demographics. Answers ranged from "Never heard of it" to statements such as "nanoparticles," "1×10^{-9} scale," "lab on a chip," "space advantages (small circuits, size reduction)," "nano-toxicity concerns," and "very small things, nano tubes, carbon nanotubes, nano particles, gold, thin films, lots of potential for new products, but also uncertainty about releases of these materials into the environment (water, air, human lungs, etc.)."

We organized these responses into three categories: high understanding, defined as a reasonably precise understanding of scale plus unique applications or implications; scalar understanding, defined as an understanding of only the scale of nanotechnology, most often as incredibly tiny; and low understanding, characterized by statements of no knowledge and highly inaccurate guesses. As shown in Fig. 2.1, well over half of respondents (59%) displayed at least a general awareness of nanotechnology's scalar definition, as something very tiny and potentially very important. However, only 9% of the respondents demonstrated a high level of understanding, including having multiple applications, most commonly noted being in information technology. Only 32% of respondents displayed little to no understanding.

While looking at variations by region or type of civic association, we must place a number of caveats on the results, as they cannot support

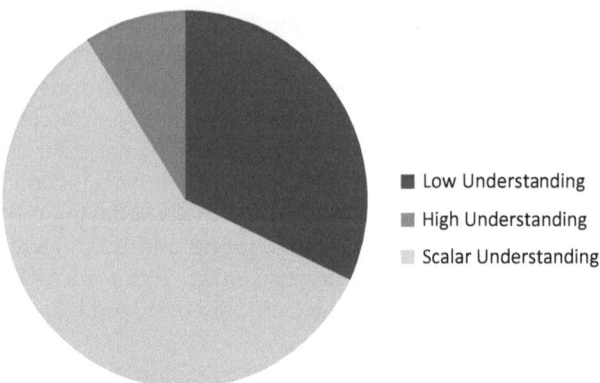

Fig. 2.1 Levels of public understanding

conclusions about regionally distributed knowledge or about knowledge distribution among group types. Because our research was limited to 11 groups around the nation, we do not have enough information here to generalize, especially given that within some regions or group types only one or two of the cooperating associations completed the surveys. In particular, the presence of only one religious association in these results may mean an unusually high knowledge and awareness in that specific group, which may not be broadly indicative of any difference found in religious groups more generally (Figs. 2.2 and 2.3).

Nonetheless, we do see in these numbers a dynamic in which groups and regions have a range of knowledge, with at least a common tendency for many members, if not the majority, to enter into the event already understanding the basics of the scalar definition of nanotechnology, even if they cannot fully define it or do not grasp its transformative potential. Likewise, in two-thirds of the groups we studied, at least two members displayed a high initial understanding of nanotechnology, with one third having no members whose survey responses indicated such an understanding. In only two of the groups did we find members with no understanding outnumbering those with at least a scalar understanding. In the most knowledgeable group, 43% of the respondents showed high understanding, 50% showed a scalar understanding, and 7% displayed low understanding. In the least knowledgeable group, no one showed high understanding, 25% showed scalar understanding, and 75% showed low understanding.

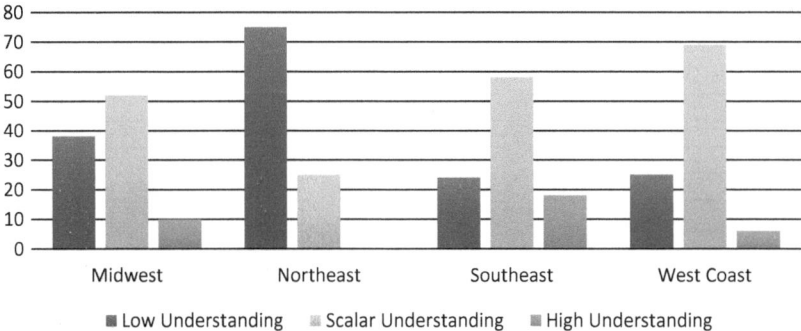

Fig. 2.2 Distribution of understanding by region

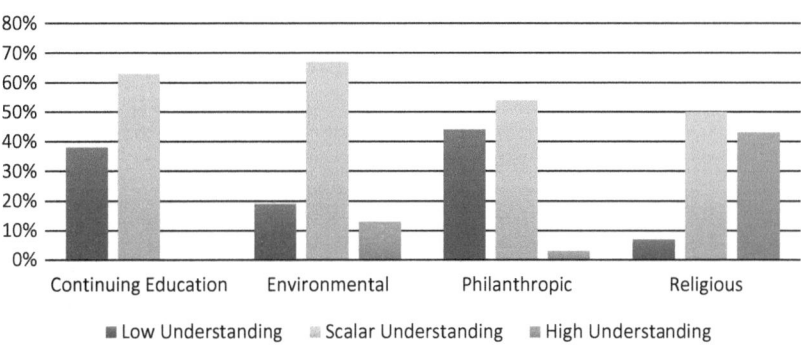

Fig. 2.3 Distribution of understanding by group type

These numbers did not tend to cluster around the average formal education level for each group or around most other demographics. In education, no discernible pattern emerges, while in income we see more indication of an inverse relationship than anything else (Figs. 2.4 and 2.5).

Again, the conclusion to draw from this data is that there tend to be wide variations in group knowledge about nanotechnology that will not be revealed by the standard categories in which we classify elements of civil society (in this case, income and education level). While previous research has indicated an antagonism between religious association and emerging technologies such as nanotechnology (Brossard et al. 2009), in our study, the religious organization was among the most well prepared

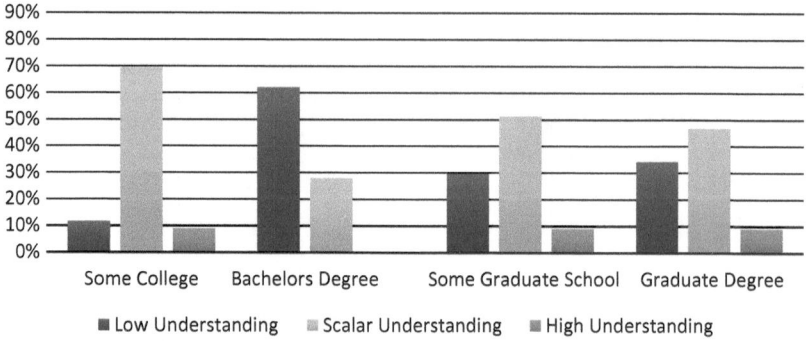

Fig. 2.4 Distribution of understanding by education level

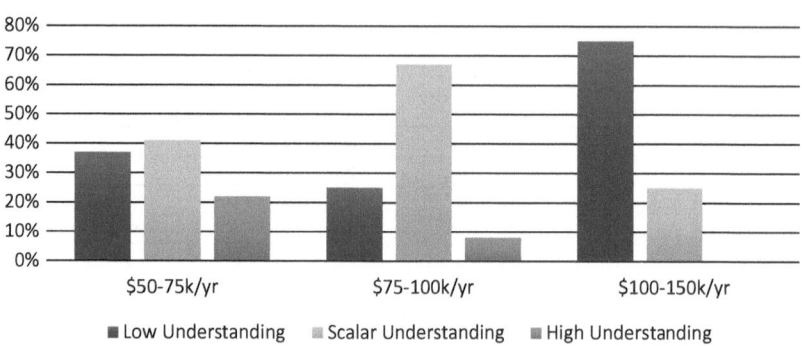

Fig. 2.5 Distribution of understanding by income level

and informed of the groups. Likewise, while one might expect higher levels of formal education to correlate to higher understanding, such an expectation would be misleading to anyone planning to speak to the groups studied in this research. If one were to make any generalizations based on this survey data, they might be these: (1) more likely than not, most audiences, even small ones, will contain a couple of people with a decent grasp of what nanotechnology is, and (2) inversely, while more rare, even a well-educated group in a privileged socioeconomic demographic can potentially include only a few members with even a basic scalar understanding of nanotechnology.

Our observations of the events and audience responses further complicate this dynamic. As speakers invited by these groups tended to begin with the most rudimentary explanations of nanotechnology, some asked audiences if they knew what the term meant. In one case, a speaker who clearly had presumed the audience was ignorant asked what "nano" meant, and one member replied, "10 to the negative 9th!" Either because she did not hear the response or because she had prepared so thoroughly to speak to an audience with no knowledge, she did not acknowledge the reply and continued as if no one knew. During another event, when the speaker did not provide the 10^{-9} definition, an audience member specifically asked, "Is a nanometer 10 to the minus 9?" Such questions indicate at least a precise scalar awareness on the part of some audience members and a willingness to engage in discussion on the topic.

In 8 of the 11 events, we saw further indication that some members of these groups arrived to events with a higher-than-average understanding of the technology and its related principles, as well as occasional indications that they were able and willing to engage in discussion. The three exceptions—the groups in which coders did not see evidence of significant prior knowledge among audience members—were a philanthropic group in the Northeast, an environmental group in the Midwest, and an environmental group on the West coast. This is not to say that these audiences did not bring some significant prior knowledge, only that their comments and questions were more responsive to subjects initiated by the speakers and, hence, could not provide evidence of such prior knowledge. In the other eight events, we saw comments and questions that clearly demonstrated knowledge brought to the event by audience members, not derived from the presentations or information provided by the speaker(s).

As with the survey data, our observations indicate a wide variation in the level and types of knowledge that audiences bring to events. Real audiences found in public groups, those voluntary associations in which people tend to form opinions and generate collective action, are likely to include members who have some prior knowledge that they are able and willing to share with fellow members. While certain elements of the "science-fiction" edge of nanotechnology (e.g., self-assembly) do appear in public discussions, those topics were the exception in what audience members brought to the observed conversations and tended not to generate further discussion or follow-up questions.

Combining the survey data with the observational data, we find that more than two-thirds of these groups had active members with enough basic knowledge to allow the groups to engage in meaningful conversation about nanotechnology, likely without the aid of expert speakers. Likewise, we found that we could not determine the collective knowledge and capacities of the groups through any demographic characteristic. Instead, any such determination would require specific questioning of the group members about nanotechnology itself. One group in particular reinforced these conclusions by not including an expert speaker and conducting its event based upon members' own education and research. While the members initially sought a speaker from a specific national organization, when that speaker was unavailable they chose instead to educate themselves, organize and prepare their own event, make a presentation themselves (four members of the group spoke), and conduct the discussion period at the end.

The contact person for this organization initially was hesitant about doing an event on nanotechnology because she had not heard much about the topic. However, after she and some fellow members read what they found through their own research, they decided not only that it was something highly relevant to their association but also that they had sufficient information to bring together their own presentation. They made handouts and slides, brought in product examples, and produced original summary material for the audience, all with a focused concern for the applications and issues that would be most salient to their members. I have chosen to call this approach auto-engagement, reflecting that the members not only engaged themselves but also did so through an autodidactic model.

Based upon this research, we have every reason to expect that about two-thirds of the civic associations across the country would have the existing knowledge and information resources within their own members and connected communities to inform themselves about and have meaningful discussions of nanotechnology if the issue should become relevant to them. Though this leaves a large portion of civic associations without that vital knowledge base, we should not underestimate the capacities present in the majority of such organizations. What keeps groups from bringing these capacities to bear on nanotechnology appears to be a disconnect between the topic and their particular goals, interests, and commitments. Once such connections are made—once they find reasons to be interested—most organizations will be able to inform themselves collectively and come to shared opinions about the topic. If the history of the role of existing public groups in the formation of social movements and political action is any guide, then we should fully expect that these organizations will take the lead if or when a crisis or rising concern about nanotechnology catches their attention (this may include crises or concerns about the lack of significant development as well about regulation or health and environmental risks).

One final implication of this dimension of this study concerns the availability and ease of use of public information about nanotechnology. When a civic association's members begin research, they are likely to set out on the Internet and/or seek information through existing and trusted channels (as was the case with our experience). Simply put, this means the two most significant procedures that any organization can follow to aid in intelligent and useful public responses to emerging technologies are, first, to build attractive, publicly available, and highly deployable sources of reliable information, and, second, to establish relationships and connections between the organizations providing that information and all dimensions of civil society so as to build trust and mutual respect. Prior opinion of information sources played a major role in information selection by the one association that took an auto-engagement approach. Similar prior relationships affected the selection of experts to speak to the organization in a number of additional cases. Such relationships will not be built ad hoc, but must emerge from months, if not years, of work.

Additionally, we should not understate the implications of these findings for future work in public engagement with emerging technologies. Public engagement research that purports to represent public opinion and priorities in development and regulation of new technologies must strive to accurately depict how existing publics actually interact with, understand, and come to opinion about technology. It must learn to listen to existing public groups, as they exist "in the wild." Doing so reveals that public understanding of nanotechnology is much more complex than survey research depicts, and that complexity has practical implications for public engagement, research on public understanding, and on science communication.

Misunderstanding and misrepresenting publics' capacities and knowledge means feeding bad data into our "upstream" engagement efforts, leading to bad decision making and policies that claim to reflect public sentiment but better reflect researcher bias. Likewise, such research and the skepticism it generates about public engagement, tends to produce ineffective and even counterproductive science communication. Consider the public event mentioned previously in which a scientist ignored an audience member who replied "10^{-9}" when she asked for a definition for "nano." How engaged do you think that member of the public felt after being ignored by the "expert" guest? Given that he was a respected long-time member of that public group, how likely is it that other members of the group noticed how he was slighted? It would be highly unusual for such an instance to pass without implications for that speaker's credibility and standing with the audience.

In a similar case, a continuing education group had invited two "expert" speakers to present on nanotechnology to their audience of adult learners. This was an event hosted by a local community college, the speakers were no doubt qualified and credible speakers on nanotechnology and science, and the audience was made up of local community members (many of whom regularly attended such continuing education events at the college). Much of the audience appeared to be over 60 years old (and our pre-event survey indicated this audience skewed significantly older than the norm, though we did not share that information with the speakers). The speakers had prepared an impressive set of slides and had a robust amount of information, but were clearly prepared for an audience who would know little to nothing about nanotechnology.

Not 10 minutes into the presentation, this assumption was revealed as folly. The community college just happened to sit near a major scientific research facility, one of the top in the country for high-level work in physics and related fields. As a result, the community housed not only a large number of active scientists, but also an unusually high concentration of retirees with advanced degrees in science and decades of experience working as researchers. Based on the interactions that made up the next hour or so, it appeared that perhaps as much as a third of the audience had a level of understanding of physics and science that made it challenging for the "expert" presenters to keep up. Yet, unfortunately, these two presenters did what far too many science communicators do when challenged: they tried to keep control and retain authority. They tried to derail conversations driven by the public group and drive the audience back to their slides and their prepared presentation, all while having to dodge questions they could not answer, pleading that their own expertise was in a different area than what the audience wanted to discuss. The result was not only two clearly shaken "expert" speakers but also an audience that was visibly uncomfortable with having to sit quietly while being presented with information that raised important questions about fundamental scientific principles. In this case, even pre-event survey data would not have been helpful to the expert speakers. As seen in Fig. 2.3 above, the pre-event survey of the continuing education group did not reveal that group members possessed high levels of understanding. This may be further evidence of the weakness of survey methods in assessing public understanding of science.

In both these cases, the speaker who ignored the audience member and the speakers who tried to direct the audience back to their prepared slides, we see what communication researchers call monologic communication. Monologic communication uses the metaphor of the monologue to describe a style and attitude of communication where speakers talk at each other, often displaying low levels of respect, difficulty sharing power, and ineffective turn taking. Communication researchers interested in the effects of monologic communication often turn to the work of Martin Buber to define the terms. Buber classified three attitudes toward communicating with another person: monologue, technical dialogue, and dialogue. As Arnett et al. (2009) summarize, "Monologue looks to the self for answers, and technical dialogue looks to public feedback" (p. 82). In both these cases, we see the "experts" look to themselves for the answers, which is understandable. After all, they were invited to speak and occupy the position of expert. Having been positioned as the expert, they likely feel pressure to be the source of the answers and information, and to control the conversation. Still too much public engagement with science and technology take this form, perhaps because many of the parties involved (scientists, regulators, and even public members) expect the "experts" to be the source of the answers.

Meanwhile, technical dialogue, looking to public feedback, better represents many of the current trends toward "upstream" engagement and some of the move toward involving citizen-science in public engagement efforts. As James Honeycutt (2011) put it, "technical dialogue is prompted by the need of objective understanding" (p. 196). In the context of public engagement, technical dialogue seeks to involve all the participants in producing a kind of understanding, an understanding of the topic or issue discussed and perhaps of the existing opinions and knowledge that the people at the engagement event wish to share. It seeks input or feedback, usually with an eye to more accurately capturing more objective forms of knowledge and data, often with a focus on the topic of discussion (as opposed to understanding the interests, motivations, cultures, and histories of the groups being engaged). Public engagement events where speakers engage in technical dialogue produce significantly better outcomes than those using a monologic approach. Technical dialogue is capable of producing more accurate understandings and more reliable data than monologue. It also does provide some limited empowerment to public groups, as some of their input can affect the conclusions and beliefs about the topic of the engagement. In the case of "upstream" engagement, technical dialogue has been very effective at breaking the dominance of monologue as the model for both science communication and science policy.

For all its value, though, technical dialogue has its limitations and leaves important information behind. Because technical dialogue and its corresponding idea of "upstream" engagement still tend to demarcate the expert and nonexpert in traditional ways, such engagement events tend to only scratch the surface of publics' capacities and often can glean only a superficial view of their values, priorities, cultures, and motivations. Moving beyond technical dialogue to dialogue (in Buber's terms) would require destabilizing the line of demarcation between expert and nonexpert, as well as introducing opportunities for all the participants to set the agenda and direct the conversation at the event. Buber distinguishes between forms of communication that treat the interlocutor as an "it" and those that treat the interlocutor as a "thou" (Buber 2004/1923). Such "I-It" relationships value the other person primarily, if not solely, based upon their utility for my own values, needs, and purposes. In short, the other person is an "it" that I desire to use, mobilize, or exploit. Most certainly, monologic communication reflects this "I-It" relationship, as speakers see audiences as passive receivers of their "expert" knowledge or holding flawed opinions that the speaker has come to correct. Technical dialogue may also be guilty of this kind of "I-It" behavior, as it often treats the other primarily as a tool for the speaker's intended desire to reach a more accurate objective understanding of the subject being discussed, or as a source for simple feedback or public comment on a policy or proposal.

Dialogue, Buber argues, requires a more open and multilateral form of communication, in which the stakes, objectives, and values are open for all the participants to negotiate. He calls this an "I-Thou" relationship. One of the markers of such and "I-Thou" dialogue is the willingness for the speaker to put her or his own ideas, goals, priorities, and values at risk by inviting the other, in this case the audience, onto a level playing field, where all can converse as equals with equal right to question, to speak, and to educate one another. Immanuel Kant's moral philosophy is grounded upon a similar notion, to treat others always as an end in themselves and never as a means only. In the context of public engagement with science, this would require surrendering the authority of the "expert" position, recognizing that expertise is negotiable and distributed, and opening oneself to challenging beliefs and values. If we could see such a move in public engagement with science, we might be able to call such events dialogic engagement. Such events would be a dramatic departure from the vast majority of public engagement with science and technology events that have been held in the United States and abroad.

Two public engagement events in this study arguably demonstrated this more dialogic approach and demonstrate the promise of dialogic engagement. First, the one religious association in this group organized their event in what appeared to be a relatively traditional manner, with a speaker who had a set of slides and prepared presentation, followed by a discussion. Yet, as the speaker presented, the conversation opened up, apparently naturally. The speaker's presentation was sidelined to make space for an organically emerging conversation. That conversation was not simply a back and forth between the speaker and the audience, but a free-flowing discussion among many of the attendees, with significant periods in which the invited "expert" speaker was silent and the "audience" spoke to each other at length.

One notable such moment was a discussion of the relationship between size and surface area. A couple of audience members were unsure how to make sense of the claim that as size decreases, surface area increases. This is a very important concept for many nanoparticles and their unique capabilities. Understandably, it can seem counterintuitive to say that diminishing size increases surface area, yet this is true in ways that matter deeply in physics, chemistry, and related fields. When the question emerged, the "expert" speaker was having some difficulty finding an accessible way to explain the concept and was not effectively hearing the person's reasoning behind the question. Other members of the audience actually took up an explanation and not the "expert" speaker, engaging in a sustained conversation of approximately 10 minutes, involving examples and even an ad hoc visual aid. In the end, the confusion was found to be cause by an imprecision in the statement, which when clarified made the issue clearer but also helped reveal how the claim is, in one sense, not true. If one has a solid cube measuring 24 in in each dimension, it will have more total surface area (3456 in^2) than a solid cube that is 2 in in each dimension (24 in^2). So, in this sense diminishing size means diminishing surface area, not increasing surface area. However, the statement is true in a different sense. The 2-inch solid cube has a surface area that is larger in proportion to its total mass than the surface area of a 24-inch solid cube. The 24-inch cube has 3456 in^2 of surface area ($6*24^2$) and $13,824$ in^3 of mass (24^3). The ratio of surface area to mass is 1:4. On the other hand, the 2-inch cube has 24 in^2 of surface area ($6*2^2$) and 8 in^3 of mass (2^3). The ratio of surface area to volume is 1:1.33. This kind of technical explanation may be fine for some, but what the audience members did to explain it to each other was a much simpler and more intuitive way to demonstrate the principle without any numbers or math. Their explanation went something like this: imagine you have a cube, it has six sides, each of those is the surface area, and the inside of the cube is the mass. Now cut the cube in half and make two smaller cubes. Now you have the same amount of mass, but two extra sides or faces (where you cut it in half), increasing the surface area. If you cut each of those in half, you still have the same mass but have added two more surfaces, again increasing the surface area. Some savvy scientists and teachers demonstrate this with a stack of 24 standard six-sided dice, but, in this case, the audience members used makeshift materials they found in the room to the same effect. In the end, the audience members' discussion among themselves better accomplished the monologic and technical dialogic goals of the "expert" speaker.

What made this moment particularly dialogic, however, was not so much in what they discussed but in how they engaged one another. Because these audience members had a long-standing relationship as fellow members of a religious congregation, they valued each other's feelings and respected one another. As a result, rather than simply explain the principle or show how the question was wrong, the fellow audience members could demonstrate this important principle of physics and its implications for nanoparticles while also vindicating the question by noting the imprecision in the shorthand statement that "as size diminishes, surface area increases." Doing so put high value on the other person and on the relationship between the interlocutors, treating them as "Thou" rather than "it" and making a space for everyone to be valued and understood as intelligent, sensible, and responsible participants. The other audience member and her or his contribution were valued and affirmed, which only strengthened the commitment to scientific accuracy and precision.

Likewise, we see dialogic communication in the group that chose a form of auto-engagement, where four members of the group researched and presented on nanotechnology to their larger community. In this case, all the presenters and attendees were, essentially, in-group members, and no outside speaker or "expert" had been brought in to speak. The result was, first, that the audience showed little doubt as to the authenticity and goodwill of the four speakers. They also were comfortable asking questions and sharing their own information, ideas, and opinions. The speakers, meanwhile, clearly brought a great deal of new information to the audience, but were not saddled with the expectation of playing the "expert" role. Unlike in the case of the two experts in the continuing education program, in the auto-engagement event a challenging question or a gap in knowledge was simply an opportunity for the members of the group to reason together or for a member of the audience to share her or his own information.

Even when an audience member shared information that was viewed by others as flawed or unreliably, it was handled with a care that reflected the importance of maintaining the positive relationships between the group members. For example, one audience member very briefly spoke about the risks that nanorobots could self-replicate, even spreading out of control like a cancer. Commonly called the "gray goo" scenario, most consider that a science-fiction plot. Not only would it require a level of advancement in nanotechnology that appears to be very far away, but even if we reach that highly advanced stage, the risk is considered to be easily manageable. Most of the "expert" speakers we studied would have corrected this audience member and felt it necessary to show her fear as unwarranted. In my opinion, this would have been a disastrous approach because the audience member's high tone and excited speech showed that she was performing for her fellow group members. She wanted, as most of us do at times, to demonstrate her knowledge and value to her peers. To invalidate her explicitly could be disastrous.

Of course, as fellow members of the group and having an established relationship with this audience member, the situation was not lost on the speakers. Rather than shut her down or correct her, they chose a passive validation that quickly moved off topic. They replied that the comment was interesting, they did not know about such a scenario, and had not seen it in any of their research, then quickly moved on to the next audience member who wanted to speak. Note that unlike the scientist who ignored (or did not hear) the audience member offer "10^{-9}" as a definition of nano, these speakers validated the audience member, allowed her to save face, but passively undermined the legitimacy of her concern by noting it was not in their own research. Moving on quickly to the next question also communicated that this was not a serious concern, but without having to directly negate the speaker. That kind of care for the other person and the relationship is part of what made that encounter more dialogic without having to open the floor to extended discussion of something that would be largely irrelevant to most of the group (as well as largely discredited).

One may notice that both of the examples I have given for more dialogic engagement highlight the communication between audience members. Being a member of the group or public you are speaking to is not necessary to be more dialogic, but it does make it easier. In fact, when we speak with people outside the groups and publics we participate in, a dialogic approach might be even more important, but it is also more challenging. Treating another person as a "Thou" and showing that you value them in themselves and not just as a means is much easier when we know the person and have a relationship to them. At minimum, it helps if we see ourselves as equals and accept that the communication event is open for everyone to learn, grow, contribute, and change. Yet, to accomplish this, we probably need to do away with the idea that engagement is "downstream" or "upstream," which imply monologue and technical dialogue, respectively. Instead, we should understand ourselves as all on the same level, without a down and an up, engaged in a web of relationships and communication events. Such a dialogic mode of communication recognizes not only that we are all "experts" in some ways and inexpert in others, but that all of our values, insights, and ideas are worthy of at least being spoken, even if the group or public decides it is not an issue of concern or even if it is deemed wrong. We value our relationships and communities and the individuals who comprise them because they make any kind of communication and collective action possible.

Finally, I want to highlight the ways in which these dialogic groups demonstrated the principle of synergy. In one case, collectively a group of informed "lay people" may produce a better explanation of a difficult concept than the expert. In another, a group of four "lay people" researched and presented information about nanoparticles in consumer goods that was arguably more appropriate to their audience and better communicated than what any of the outside experts did in other events. I contend that the core difference that made these synergies possible and made these events such successes is that, in both cases, a collective group worked cooperatively, with strong mutual respect and placing a high value on the relationship. It was not the expert versus inexpert that made the dialogic difference, but, instead, the kind of respect, care, and cooperation that comes from being a group member. The challenge for science communicators is how to likewise practice those principles even when not members of the audience or group they are communicating with. Chapter 5 goes into some detail about the communication techniques that can help science communicators accomplish this difficult goal.

One common thread connected all of the events I have discussed in this chapter and, in fact, all 11 public engagement events in this study. That is, the groups explicitly discussed the risks and benefits of different applications of nanotechnology. In doing so, they shared their values and their priorities as publics, revealing how they reasoned about risk and the kinds of applications they might most support. The next chapter goes into detail about these priorities and values, providing a detailed analysis of publics' perceptions of nanotechnology risks and applications.

WORKS CITED

Arnett, R. C., Fritz, J. M. H., & Bell, L. M. (2009). *Communication ethics literacy: Dialogue and difference.* Thousand Oaks: SAGE.

Brossard, D., Scheufele, D. A., Kim, E., & Lewenstein, B. V. (2009). Religiosity as a perceptual filter: Examining processes of opinion formation about nanotechnology. *Public Understanding of Science, 18*(5), 546–558.

Buber, M. (1923/2004). *I and thou* (R. G. Smith, Trans.). London: Continuum.

Cobb, M. D., & Macoubrie, J. (2004). Public perceptions about nanotechnology: Risks, benefits and trust. *Journal of Nanoparticle Research, 6*(4), 295–405.

Hart Research Associates. (2007). *Awareness of and attitudes toward nanotechnology and federal regulatory agencies: A report of findings based on a national survey among adults.* Retrieved from http://www.nanotechproject.org/process/files/5888/hart_nanopoll_2007.pdf

Honeycutt, J. M. (2011). Dialogue theory and imagined interactions. In J. M. Honeycutt (Ed.), *Imagine that: Studies in imagined interactions* (pp. 193–204). Cresskill: Hampton.

Johnson, D. (2009, January 14). *Nanotechnology and public engagement: Is there a benefit?* [Blog post]. Retrieved from http://spectrum.ieee.org/tech-talk/semiconductors/devices/nanotechnology_and_public_enga

Macoubrie, J. (2006). Nanotechnology: Public concerns, reasoning, and trust in government. *Public Understanding of Science, 15*(2), 221–241.

Miller, G. (2008). Nanotechnology and the public interest: Repeating the mistakes of GM foods? *International Journal of Technology Transfer and Commercialization, 7*(2–3), 274–280.

Scheufele, D. A., & Lewenstein, B. V. (2005). The public and nanotechnology: How citizens make sense of emerging technologies. *Journal of Nanoparticle Research, 7*(6), 659–667.

CHAPTER 3

Nanotechnology Applications and Risks: Valences and Ambiguities

Abstract Regardless of the amount of information public groups possess or demonstrate, most seem willing to communicate preferences and apprehensions about new and emerging technologies. One of the goals of this study was to discover such preferences and apprehensions about nanotechnologies in various public groups. This chapter details those findings as well as what the expressed preferences teach us about how public groups form opinions about emerging technologies. Few studies have mined into the details of how different applications of nanotechnology evoke different responses from various public groups. In this study, we found such data emerged naturally in the course of the public engagement events. Our results indicate that some areas of nanotechnology have broad public support, while some almost universally produce public demands for regulation.

Keywords Nanotechnology • Risk perception • Applications • Public opinion

Regardless of the amount of information public groups possess or demonstrate, most seem willing to communicate preferences and apprehensions about new and emerging technologies. This is a common element found in studies of public perceptions of new technologies, including nanotechnology (e.g., Cobb and Macoubrie 2004, p. 8). One of the goals of this study was to discover such preferences and apprehensions about nanotechnologies in various public groups. This chapter details those findings as well as what the expressed preferences teach us about how public groups form opinions about emerging technologies.

Previous studies might lead us to believe that public sentiment about an emerging technology would be positive. In general, researchers find that members of the American public are optimistic about technology, including new technologies (Einsiedel 2005). That general technology optimism led Gaskell et al. (2005) to argue that such "pro-technology cultural values" inspire Americans to be much more optimistic about nanotechnology than Europeans (p. 81). As this research might lead us to expect, studies that look at public sentiment about nanotechnology in general find a largely positive public response. Bainbridge (2002) found that 57.5% of respondents agreed that human beings would benefit greatly from nanotechnology. He found science-attentive members of the public "very enthusiastic about nanotechnology" (p. 569) and that many respondents expressed "unconditional confidence" that nanotechnologies will "benefit mankind" (pp. 566–567). Bainbridge noted, however, that most respondents did not express confidence in any specific application or technology, but instead spoke vaguely about nanotechnology's promise. This kind of vague positivity in public sentiment about nanotechnology does not necessarily translate into positive sentiment for specific applications.

Few studies have mined into the details of how different applications of nanotechnology evoke different responses from various public groups. In this study, we found such data emerged naturally in the course of the public engagement events, allowing us to code public expressions of sentiment by specific application. This means that we could break past the general positivity and gauge how publics might respond to information or even controversies about specific nanotechnology applications. The utility of such insight for industry and government should not be underestimated. Our results indicate that some areas of nanotechnology have broad public support, while some almost universally produce public demands for regulation.

The reliable codes from our study include 13 potential applications of nanomaterials that were most commonly mentioned at the 11 events. We assessed audience concern by examining only instances in which conversations included both mention of one or more of those 13 applications and some form of positive or negative response from the audience. Within the scope of audience response, we included verbal statements, the tone or disposition of questions, and nonverbal or paralinguistic responses (such as audience nodding in agreement or making audible sounds of assent or dissent). Our analysis of expressions of concern or interest (both positive and negative) among audience members indicates that 9 of the 13 applications were of the highest concern for these public groups. We then took the previously assessed positive and negative disposition codes and compared them to each instance of audience response. As shown in Table 3.1, this allowed us to produce a positive and negative disposition score for each application and to break out specific subtopics for each of those concerns.

By calculating the total positive and total negative scores for each application and then comparing the results as ratios, we produced a disposition score for each application. Disposition scores ranged from 1 (highly favorable) to −1 (highly unfavorable). Based upon audience feedback during the events, these public groups produced the following collective dispositions toward each application (Table 3.2).

Likewise, based upon audience members' most common expressions of concern, we compiled a list of motivating factors behind these scores and their relative weight based on frequency of occurrence. I am cautious to note that these relative weights are most accurately understood in relation to their specific applications; nonetheless, I believe they may also have some independent value as indicators of the types of concerns most likely to motivate publics. We calculated three scores for each of the major factors that groups articulated as the basis for positive or negative valuation: the raw score of instances of that basis, the percentage of the total articulated bases for the positive and negative valuations, and the weighted percentage based only on those bases not classified as other/miscellaneous (Table 3.3).

Based on this data, I conclude that the applications producing the most significant positive responses were in the areas of medical sensing technologies, enhancements in energy efficiency, and solar power. Medical sensing's positive appeal was naturally associated with saving human lives, as benefits of early detection have been well-established in the public consciousness. This also resonated with previous findings by Cobb and Macoubrie (2004) that "new ways to detect and treat human diseases" evoke overwhelming positive response (p. 8).

An interesting dilemma in public reception of medical applications emerges when comparing the strongly positive attitudes to medical sensing applications with the ambivalent attitudes expressed toward cancer treatment, which produced a net negative disposition score, albeit a very

Table 3.1 Positive and negative audience responses to applications

Application	Positive					Negative					
	Econ. Compet.	Save lives	Energy crisis	Food	Misc./Other	Unregulated	Insuffic. Research	Needs labeling	Long-term health effects	Human Migration	Misc./Other
Computers & smartphones	1				1	2					4
Cancer treatment		2					1				2
Medical sensing		3									
Energy efficiency	1		2								
Sensors (Generic)				3		1				1	1
Solar power	3		1		1						
Sunscreens & cosmetics					1	2	2	6	1		1
Textiles & fabrics		3			2	1	2		1	1	
Water purification & desalination					2		1		1	1	1

Table 3.2 Collective disposition by application

Application	Total positive	Total negative	Disposition
Computers & smartphones	2	6	−0.5
Cancer treatment	2	3	−0.2
Medical sensing	3	0	1
Energy efficiency	3	0	1
Sensors (Generic)	3	3	0
Solar power	5	1	0.67
Sunscreens & cosmetics	1	12	−0.85
Textiles & fabrics	5	7	−0.17
Water purification & desalination	2	3	−0.2

Table 3.3 Weighted factors influencing disposition

	Total	Percent	Wtd. Percent
Positive			
Economic Competition	5	19%	26%
Save Lives	8	31%	42%
Energy Crisis	3	12%	16%
Food	3	12%	16%
Misc./Other	7	27%	n/a
Negative			
Unregulated	6	17%	23%
Insuffic. Research	5	14%	19%
Needs Labeling	6	17%	23%
Long-Term Effects	4	11%	15%
Human Health	2	6%	8%
Migration	3	9%	12%
Misc./Other	9	26%	n/a

weak one. In part, this reflects a skepticism about claims of new cures for cancer and cancer therapies based on an expressed public perception of decades of overstated promises of forthcoming cures. Many public groups expressed frustration, feeling that they have been told they were on the brink of the next great cancer breakthrough for over 30 years. Even though cancer treatment has improved remarkably in that time, the public perception is that the big breakthrough never comes through, always falling short of its promise. In some ways, I see this as an extension and specific instance of Macoubrie's (2006) findings that people rely on past experiences and examples when forming opinion about new technologies (p. 228).

Among energy applications, we found strongly positive dispositions toward the use of nanomaterials to enhance energy efficiency and to improve solar electricity generation. Two primary benefits drove this perception, the first being the obvious role of these technologies in ameliorating either a current or looming energy crisis. The second, less obvious but still very strong, was the role such nanotechnologies could play in economic competitiveness, particularly the United States' position in the global economy. Concerns about competition from China, specifically, and other countries, more generally, motivated a large number of audience members to express support for nanotechnology's potential in solar power technology. In fact, audience members were three times more likely to respond positively to solar power for reasons of economic competitiveness than for any other reason (including environmental reasons).

Inversely, the most significant negative responses to nanotechnology, and specifically to products containing nanoparticles, are focused on the perceived failure of any agency to effectively regulate, sufficiently test, or provide sound labeling for cosmetics and sunscreens. Cosmetics and sunscreens produced negative responses at nearly twice the rate of any other application and evoked only one countervailing positive audience reaction. I discuss this case at some length in Chap. 4 to make the case that there is no other application in more dire need of improved public communication and more ripe for increased governmental regulation.

Audiences' negative dispositions clustered around issues of labeling, specifically the lack of labeling and clarity of labeling. This was amplified by related concerns about insufficient research on effects of nanoparticles, inadequate regulation of the use of those particles, and human health and safety risks. This particular convergence produced an especially negative disposition toward brands that have significant connections to nanomaterials.

For example, when discussing the difficulty of making an informed choice as a consumer of cosmetics and sunscreens, one public group turned to a strategy of arguing that whole brands be considered either nano-free or likely to use nanoparticles. Of the brands discussed at the event, L'Oréal was the one most strongly associated with the widespread use of nanoparticles, but other mainstream major brands (e.g., Revlon) were also associated with the likely use of unlabeled or undeterminable nanomaterials. Audiences also expressed negative attitudes toward specific brands normally perceived as more natural that they believed were likely to contain nanoparticles (such as products from The Body Shop). I have no specific knowledge of whether or not these brands use nanoscale ingredients, nanoparticles, or nanotechnologies, and in most cases, audience members and "expert" speakers were uncertain about what brands might use what nanoparticles in what products. That ambiguity was precisely what produced the public sentiment that if you wish to avoid exposure to nanoscale ingredients in cosmetics, the wisest option is to avoid the entire product line of any suspect or uncertain brand.

Marketing terminology used in cosmetics also tended to reinforce audience concerns about health and safety issues and their desire for labeling. As cosmetics marketing emphasizes products' capacities to penetrate "deep" into the skin, to affect the health and condition of the skin, to "nourish," and similarly to affect the body, audiences find them at odds with assurances that the nanoparticles do not penetrate the skin and do not interact with the body. Such claims elicited some dubious responses not only from audiences, but also from "expert" speakers who questioned the assurances when products either by design or by common use would come into contact with broken skin (as with lotions, sunscreens, and many cosmetic products), may be inhaled during application (as with dry powder mineral makeup), or could be ingested via transfer from the hands or lips (most clearly with lipstick but also with lotions and sunscreens).

One of the reasons that cosmetics and sunscreens fared so poorly as a category is because public groups tended to weigh the exposure to nanoparticles and potential risks of such exposure against their perception of the benefits and value of the application. In a case like medical sensing technology, public groups would expect exposure to be rare and the potential benefit to be significant to their individual health. With an application like solar power generation most public groups would not expect any direct exposure but would still hope to accrue the economic benefits of having domestic corporations leading the alternative energy market. However, in the case of cosmetics and sunscreens, public groups are likely to perceive exposure as both frequent and widespread.

In this study we found that the negative sentiment toward nanoparticles in cosmetics and sunscreens emerged both as expressions of perceived risk and as indications that public groups did not perceive these as high-benefit applications. It may be especially difficult to justify individual or societal benefits to including nanoparticles in most cosmetics, which are already perceived as products with little societal value. However, there is little public doubt about the societal and individual benefits of sunscreens. In the public engagement events we studied, it appears that the negative sentiment toward nanoparticles in sunscreens was due to three factors.

First, the aggregation of cosmetics and sunscreens into a single category may cause sunscreens to suffer from a reverse halo effect or devil effect (Nisbett and Wilson 1977). Second, negative public statements about both sunscreens and cosmetics often were intertwined with complaints about the difficulty in determining whether a product contains engineered nanoparticles. Consumer desire for transparency and reliable labeling was a clear factor in negative perceptions of cosmetics and sunscreens. Third, the public groups in this study did not see a significant benefit to using nanoscale ingredients in sunscreens over the other existing products. While some evidence exists that sunscreens based on nanoparticles may be safer and more effective that other options (Deng et al. 2015), this did not translate into public perception of a net benefit.

Public groups in this study clearly demonstrated a complex collaborative approach to considering risks and benefits. While previous research generally leads us to expect perceived benefits to reduce risk perception and increase tolerance for risk (Slovic et al. 1986), this study offers specific evidence of this behavior in consideration of nanotechnologies. In most cases, the groups we studied gave latitude to applications that were seen as offering unique and important advantages over alternatives. For example, in discussions of clothing and textiles, audience reactions differed when the topic was stain- or wrinkle-resistant clothes versus sheer thickening fluids containing nanoparticles used to coat clothing to protect law enforcement and prison guards against knives. The trivial advantages of stain and wrinkle resistance did not mitigate concerns about continuous direct exposure to the fabric that daily wear might produce. On the other hand, seeing how a special coating could save the life of a law enforcement officer in a way that existing options do not make practical was a clear and unique benefit of significant value. Implications of this comparative collaboration for both regulation and science communication are discussed in Chaps. 4 and 5.

Public groups in this study were consistently negotiating between competing values to assess various applications of nanotechnology. In making these weighted comparisons between risks and benefits, one factor that can easily be overlooked is the function of value hierarchies in public perceptions of risk and risk tolerance. Ever since Milton Rokeach's foundational work on value hierarchies (1973), social scientists have been aware of the role of value hierarchization in public sentiment. In cases where publics are weighing relative risks and benefits of a perceived risk, such as an emerging technology, we should expect value hierarchies to play a role in prioritizing certain benefits over others and valuing which benefits warrant more or less risk. In the case of nanoparticles used to coat clothing, a very simple and largely uncontested value hierarchy emerges: saving human lives (especially the lives of public servants in high-risk jobs) is a higher value than keeping your pants looking clean and crisp. On the other hand, in the case of solar power applications we note that the high value score was common in the public groups but due to divergent reasons. The fact that most groups were positively inclined to solar power for economic reasons while a smaller number were so inclined for environmental reasons reflects the diversity of their value hierarchies. Study of value negotiation in public groups will be essential to understanding how public groups form opinion and come to action on any topic, including emerging technologies.

Three major implications for public perceptions of risk emerge from these results: risk perception is relative to a public group's collective values, risk perception is cross-contextual, and generic tech-optimism quickly gives way to these relative and cross-contextual forces when discussing specific applications. The kinds of assessments we observed public groups making in this study were not only statements made by single individuals but, in many cases, moments of collective reasoning and argument. As with the example in Chap. 2 of the group discussing the relationship between size and surface area, groups discussed benefits and risks to come to opinion and express collective sentiment. In some cases, this was quite explicit, but in many cases, we saw simple statements of affirmation or paralinguistic cues such as murmurs of assent that affirmed one speaker's views and not another's. While not what we might consider traditional deliberation, these can be strong social forces in the establishment of group sentiment, which not only can weigh heavily on an individual's own views, but may be the primary driver of collective action within a group. This evidences again that the study of publics cannot be done merely by aggregating individual opinion or behavior, as public groups function as a collective organism, forming opinion and coming to action in synergistic ways while also finding their own social ecology establishes and enforces collective value hierarchies.

We also find that public groups tend to understand both benefits and risks cross-contextually, applying a whole raft of social, cultural, and historical elements to judge the claims of "experts" and fellow group members. While this tended to affect public groups' discussions of every topic, it was especially pronounced in discussions of cancer treatment and regulations. It was initially surprising to find that the public groups held such a positive regard for using nanotechnology in medical sensing technology yet had such a negative attitude toward cancer treatment applications. Given the comparative collaboration and value negotiation that we saw in other topics, we expected the promise of nanotechnology in cancer treatment to score quite positively for audiences. Only after viewing the specific negative statements in the videos of the public engagement events did the cause of this negative score become clear. Audience brought into the discussion a host of previous claims about the next great cancer treatment and their history of disappointing patients. Not only did that undermine essentially any future technology's promise of assisting in the fight against cancer, but it seemed to intensify the greater the claimed potential benefit. An evolution in treating a specific cancer was far more likely to gain approval than a revolutionary technology promising to cure all forms of cancer. Thus, while some have expressed concern about public perception of technologies such as genetically modified organisms impacting their perception of other emerging technologies, we found a much more complex set of influences that were grounded in that public group's own culture, history, and traditions.

Finally, I want to note that whenever a public group began discussing a specific technology, their frame of reference and method of evaluation tended to shift. General sentiments about technology or even nanotechnology and overall trends in their discussions often gave way to a specific framing of the issues or a variance in perceptions of product categories based on the specific applications within those categories. Just as there is no single entity called "the public" but, instead, a host of interacting and overlapping publics, public groups do not tend to encounter or experience a broad entity called technology or even a general entity called nanotechnology. Instead, their perceptions of risks and benefits, and their tolerance of risk in order to accrue a benefit, are formed, evaluated, discussed, and acted upon within the context of specific applications. One important implication of this fact is that studies purporting to reveal a public sentiment about a technology that only address that technology as a broad class provide little guidance and may even significantly mislead us about how publics perceive and will respond to that technology as it is deployed and encountered in the world. That is, since it will always only be deployed and encountered by publics within the context of a specific application, and since publics perceive and tolerate risk differently for different applications of the same technology, if we wish to have meaningful and actionable research on public perception of a technology, it must distinguish between different applications and study public perception of those specific applications.

This is likewise the case when it comes to regulating technologies and understanding public sentiment about regulations. The next chapter takes up this issue by discussing what public groups in this study expressed about regulation of specific applications of nanotechnology, specific forms of regulations, and different kinds of regulating bodies. This data offers significant guidance to industry and government officials concerned with responsible regulation of emerging technologies in general, and nanotechnology specifically.

WORKS CITED

Bainbridge, W. S. (2002). Public attitudes toward nanotechnology. *Journal of Nanoparticle Research, 4*(6), 561–570.

Cobb, M. D., & Macoubrie, J. (2004). Public perceptions about nanotechnology: Risks, benefits and trust. *Journal of Nanoparticle Research, 6*(4), 295–405.

Deng, Y., Ediriwickrema, A., Yang, F., Lewis, J., Girardi, M., & Saltzman, W. M. (2015). A sunblock based on bioadhesive nanoparticles. *Nature Materials, 14,* 1278–1285.

Einsiedel, E. (2005). In the public eye: The early landscape of nanotechnology among Canadian and U.S. publics. *AZoNano: Online Journal of Nanotechnology.* Retrieved from https://www.azonano.com/article.aspx?ArticleID=1468

Gaskell, G., Eyck, T. T., Jackson, J., & Veltri, G. (2005). Imagining nanotechnology: Cultural support for technological innovation in Europe and the United States. *Public Understanding of Science, 14*(1), 81–90.

Macoubrie, J. (2006). Nanotechnology: Public concerns, reasoning, and trust in government. *Public Understanding of Science, 15*(2), 221–241.

Nisbett, M. C. (2010). Framing science: A new paradigm in public engagement. In L. Kahlor & P. A. Stout (Eds.), *Communicating science: New agendas in communication* (pp. 40–67). New York: Routledge.

Nisbett, R. E., & Wilson, T. D. (1977). The halo effect: Evidence of unconscious alteration of judgments. *Journal of Personality and Social Psychology, 35*(4), 250–256.

Rokeach, M. (1973). *The nature of human values.* New York: The Free Press.

Slovic, P., Fischhoff, B., & Lichtenstein, S. (1986). The psychometric study of risk perception. In V. T. Covello, J. Menkes, & J. Mumpower (Eds.), *Risk evaluation and management* (pp. 3–24). Boston: Springer.

CHAPTER 4

Government Regulation of Nanotechnology: Imperfectly Essential

Abstract Just as this study of public groups revealed sometimes nuanced perceptions and valuations of the risks and benefits of nanotechnologies, it also uncovered a complicated network of attitudes about regulation. This chapter reports on the study's findings of public groups' attitudes about regulating nanotechnology, including perceptions of different kinds of regulatory bodies and industry self-regulation. Based on close examination of the statements and reactions in these public groups, we found ambivalence within most public groups about a wide variety of regulatory schemes, but a widespread and strong support for mandatory labeling of nanoscale ingredients and engineered nanoparticles in consumer products.

Keywords Nanotechnology • Nanoparticles • Regulation • Labeling • Consumer goods • Cosmetics • Sunscreens

Just as this study of public groups revealed sometimes nuanced perceptions and valuations of the risks and benefits of nanotechnologies, it also uncovered a complicated network of attitudes about regulation. Regulation is often a controversial topic, and regulation of new and emerging technologies can add a level of opacity and uncertainty to the debates. Regulators, legislators, and industry often seek out public opinion as a part of the regulatory process or to bolster their positions in such debates. This chapter reports on the study's findings of public groups' attitudes about regulating

© The Author(s) 2018   51
P.J. Gehrke, *Nano-Publics*,
https://doi.org/10.1007/978-3-319-69611-9_4

nanotechnology, including perceptions of different kinds of regulatory bodies and industry self-regulation. On the basis of close examination of the statements and reactions in these public groups, we found ambivalence within most public groups about a wide variety of regulatory schemes, but a widespread and strong support for mandatory labeling of nanoscale ingredients and engineered nanoparticles in consumer products.

In Chap. 3, we saw that public groups negotiate values, consider social factors that amplify risk perceptions, and weigh risks and benefits through a complex collaborative approach that includes cross-contextual evidence. We found that they use similar processes when considering regulatory options. Relative weighing of risks and benefits, cross-contextual evidence about regulatory bodies, collaborative approaches to evaluating regulatory schemes, and negotiated values all played a role in public perceptions of regulating nanotechnologies. As a corollary to the social amplification of risk principle, we could call this the social perception of regulation.

The social perception of regulation follows some norms we might intuitively expect and others consistent with the behaviors that drive the social amplification of risk. For example, as one would expect, the higher the public group's perception of the risks of an application, the more likely they were to be positively disposed toward regulating that application. Such risk perception might be heightened by many of the common factors, such as lack of control, exoticness, chronic exposure, and so on.

Perhaps less intuitive is the fact that lower benefit perceptions also tend to generate more positive attitudes toward regulation. The comments we witnessed in the 11 public engagement events included in this study showed that this relationship was due not just to the tendency to perceive low-benefit applications as being riskier, but more to the tendency of groups to weigh benefits and risks against each other when thinking about regulatory options. In fact, applications perceived as having high benefits (solar power and medical sensing) generated a much higher tolerance for risk. This was most explicit in the discussion of medical applications of nanotechnology (with the exception of cancer treatment, as discussed in Chap. 3). If we consider that medical technologies will often be understood by publics as saving or extending a life, that exposure tends to be rare, is usually made with informed choice by the patient, and occurs in a controlled medical environment, it makes sense that the social perception of regulation would be less favorable for most medical applications. As discussed in Chap. 3, cross-contextual evidence deployed in public groups' discussions led them to separate cancer treatment from other medical applications, reducing their perception of the likely benefits and producing far more ambivalence about that application.

The logical conclusion is that the applications most likely to produce the highest public support for regulation will be those that public groups perceive as high-risk and low-benefit applications. While public groups did not express this rationale, there is a good deal of logic in the principle that low-benefit, high-risk applications should encounter a unique level of scrutiny. The potential opportunity cost posed by strict regulations (stifling innovation, quashing market potential) is simply a lower cost if the benefits of that application are small. This is a logic familiar to many regulatory agencies and explicitly employed by the Food & Drug Administration's benefit-risk assessment framework when evaluating new drugs and medical devices (U.S. Food & Drug Administration 2013). We found public groups were particularly inclined to deploy this logic when discussing the regulation of applications that publics believe they might encounter directly, when they might risk chronic exposure, where they could not control that exposure, and that might increase morbidity or mortality risks. In the public groups we studied, the primary category of applications that fit those criteria was consumer goods. While food packaging generated a small amount of discussion at one event, two applications triggered public concern and demonstrated stronger public desire for regulation than any others: cosmetics and sunscreens.

In fact, as cosmetics and sunscreens were among the most negatively viewed potential applications of nanotechnology, public groups expressed overwhelmingly positive views of the regulation of nanotechnology in these products (and increased regulations of these products in general). While this study does not bear sufficient evidence to make the claim with certainty, it does give us reason to hypothesize that we would find an inverse relationship between public attitudes toward an application or technology and the regulation of that application or technology. As public perception becomes negative, the public becomes more favorably disposed toward regulation. As public perception of a technology becomes more positive, the more negatively they will view regulation. This is may not be universally true and may vary significantly across contexts, technologies, and publics.

The relationship between perceived benefit and attitude toward regulation is also asymmetrical, skewed by a presumptive bias against regulation. Most of the public groups we studied expressed a significant presumption against regulation, requiring that significant reasons or evidence for regulation be supplied before they would support pro-regulation claims. Some public groups were more negatively disposed toward regulation than others, but almost all of them had some degree of negative bias.

The source of that negative bias, as expressed by the public groups that participated in this study, is a lack of faith in any of the parties who might enact, oversee, or enforce regulations. Many previous studies have found that people, especially in the United States, have negative attitudes toward government and governmental regulatory bodies (Einsiedel 2005; Hart Research Associates 2007; Macoubrie 2006; Miller 2008). Consistent with those findings, in this study, we saw public groups express significantly more negative views than positive views about government and governmental regulation. In these pubic engagement events, we saw expressions of distrust based on two primary categories: incompetence and indifference.

On the one hand, as Table 4.1 shows, the majority of negative statements revolved around areas of government incompetence. This included concerns that regulations fall short because of conflict or confusion between different government agencies. Likewise, public groups tend to view government as too slow, only establishing regulations after it is too late and products are already in use. In one event, there were multiple instances of the group simply expressing the view that the government is incompetent without specifics or explanation. Such statements, taken together, outnumber the only significant positive category for government competence ("guards us from risk") by a ratio of nearly 3:1.

On the other hand, there was some tension in the findings as to how much public groups believe the government actually cares about them and has their best interests at heart. Explicit discussion of this topic occurred in very few of the events and the instances of positive or negative sentiments weighed about equally across those few events.

Table 4.1 Public disposition toward government

	Events	Instances
Positive		
Guards us from risk	4	7
Cares about "us"	2	2
Necessary (vs self-reg.)	3	4
Negative		
Confusion b/w agencies	4	4
Doesn't care about "us"	1	2
Doesn't guard from risk	3	7
Too slow	3	4
Incompetent	1	4
Unfunded mandates	1	1

This kind of ambivalence is built into the very structure of many American government regulatory agencies, as they are simultaneously charged with both regulating an industry and promoting that industry. Take, for example, the mission of the Department of Agriculture (USDA):

> We have a vision to provide economic opportunity through innovation, helping rural America to thrive; to promote agriculture production that better nourishes Americans while also helping feed others throughout the world; and to preserve our Nation's natural resources through conservation, restored forests, improved watersheds, and healthy private working lands. (U.S. Department of Agriculture 2017, para. 2)

While most would find the listed goals and values laudable, there can often be a tension between the promotion of the agricultural industry that occupies the first half of the statement and the preservation of natural resources that occupies the second half. Many other federal agencies have missions that can create similar conflicts, such as the Food and Drug Administration's mission to "speed innovations" in medical products and to protect the public health (U.S. Food & Drug Administration 2017, para. 3). The fact is that the structure of these agencies and their missions creates a condition in which they are charged with being both protectors of the general public and allies to business and industry. These missions may not always compete, but public groups believe these agencies exist as much to benefit industry as for the public good.

Yet, even as government and regulatory agencies fared poorly in these public engagement events, expressions of sentiment about private sector and industry were much more negative. While perceived as more competent than the government, industry was also viewed as, at best, indifferent if not outright malevolent toward publics and the general population. We found 36% more expressions of negative sentiment toward industry than we did toward government, and we found no clear or direct positive sentiments toward industry. For as much as public groups appear to view government negatively and express a presumption against regulation, they view industry far more negatively (Table 4.2).

Low trust in business leaders has been found in previous studies of public attitudes toward emerging technologies (Cobb and Macoubrie 2004). Yet, our results are especially stark. We found essentially no significant positive statements toward industry, little to no trust in industry or private sector actors, and a general sense that the public needs to be

Table 4.2 Public disposition toward industry

	Events	*Instances*
Move to market before risks are understood	5	10
Motivated by profit/greed	3	11
Treat public as "guinea pigs"	2	2
Focused on the short term	5	7

guarded from bad actors in the business sector. For example, in three of our events, there was significant and repeated emphasis on the idea that private sector actors are only motivated by greed and have no concern for public interest or health unless it somehow impacts their bottom line. In two of the events, the public groups went so far as to say that private sector actors use customers and the public in general as "guinea pigs."

At best, public groups expressed concern that private sector actors are short-sighted and move products to market without sufficient due diligence. Even if not viewed as malicious or greedy and uncaring, public groups tended to see private sector actors as generally indifferent to public safety, unless it became an immediate threat to their profitability. This was reflected in discussions of industry moving products to market before they have been sufficiently tested and concerns that industry is too short-sighted and fails to see the potential long-term implications of their actions or products. One public group did discuss the potential deterrence effect of lawsuits and bad publicity, debating the likelihood that private actors avoid bad behavior because of the potential negative impacts on their businesses. Yet, even here the distrust of industry was so strong that the group quickly dismissed this dynamic, using cross-contextual evidence of bad corporate actors who easily survived what they perceived as egregious violations of public trust (such as in the tobacco and oil industries). They also dismissed the deterrence effect of lawsuits and bad publicity because such risks are, they believe, less likely to deter startups and entrepreneurs willing to take on the higher risk in pursuit of higher profits. This latter argument was made quite explicitly in a group focused on local food issues. The members of that group responded to this idea with moderate assent and no challenges. This kind of a complex evaluation of regulations, markets, and private actors reflects how some public groups generate especially robust discussions of regulation and form opinions about regulatory schemes through collaboration and negotiation.

This combination of public distrust of government and even greater distrust of private sector actors actually led public groups back around to a desire for government regulation. Even if the government is incompetent and may not truly care about the citizens, their logic goes, it is the only guardian we have against a much worse actor, private companies. In this dynamic the government is depicted as a sort of bumbling cop who may sometimes be looking out more for himself than the community. Public groups seem to view the government as something like Chief Wiggum from the television series *The Simpsons*. Wiggum is corrupt, susceptible to bribery, generally ignorant, and overall not too bright. Yet, he is also the only real authority for the enforcement of the law and the protection of the citizens of Springfield and he knows it. He tends to lord over people, demanding they respect his position, and sometimes abusing his power. He is too dumb to be trusted with anything important, too corruptible to be trusted in a pinch, too egotistical and power hungry to be ignored or bypassed, but he is ultimately the law and the person responsible for enforcing those laws. As unreliable as he is, he does still come through from time to time and people will turn to him when they need law enforcement or the kind of help that a law enforcement officer might be in a position to provide.

On the other hand, public groups tend to view the private sector as something more akin to *The Simpsons* character Mr. Burns. Mr. Burns is a businessman, owner of the Springfield nuclear power plant (along with at least a dozen other ventures), and often occupies the role of villain or antagonist on the show. He is a billionaire (in most episodes), displays shameless greed, and tends to display no regard for the welfare of anyone except himself. Burns is smart, crafty, shifty, and not to be trusted. He is a stereotype of corporate management, particularly as depicted in American film and television. More comical than *Wall Street*'s Gordon Gecko, he is nonetheless a manifestation of a similar skepticism about major actors in the private sector. Burns is depicted as distanced from society, enclaved in his wealth and power.

The public perception of government as both incompetent and at best ambivalent toward the public good makes Chief Wiggum a stand-in for public anxieties about regulation and law enforcement in general. The public perception of industry as short-sighted, greedy, and uncaring makes Mr. Burns a model of what publics expect when we speak of industry self-regulation. Yet, the most significant problematic that emerges is not one or the other of these public attitudes, but their combination into a dynamic that makes government necessary but untrustworthy because industry is far worse. What we could call the "Wiggum-Burns dynamic" forces publics to choose between the lesser of two evils. To refuse the authority and regulation of Wiggum leaves one exposed to the worst behavior of Burns. This dynamic then leads publics to call for regulation and enforcement, even while they recognize that their appeals must be made to an actor they view as untrustworthy and incompetent. It also explains why some previous studies have found that even though publics have negative attitudes toward government, they still often express the view that those governments and their regulatory interventions are necessary to protect the public (Macoubrie and Rejeski 2005).

The Wiggum-Burns dynamic also encourages people to support regulatory structures that provide as little power to government as possible, minimizing the potential risks of government incompetence and reliance upon a government they believe to be inept and corrupt. This may explain why public groups in this study expressed more favor for regulations that do not ban products or give governments control over what can be developed or even brought to market. This is also consistent with the high value that some public groups placed on economic development and the positive relationship they saw between economic development and technological innovation. This combination of the Wiggum-Burns dynamic and prioritizing economic development may also explain why, even when discussing applications with little to no perceived benefit and high perceived risk (cosmetics), the most preferred form of government regulation was mandatory labeling. In fact, mandatory labeling of engineered nanoparticles in cosmetics receive almost unanimous support in this study.

Almost universal support for mandatory labeling of engineered nanoparticles in consumer goods is a logical outcome of a public that feels it cannot trust government or industry, the first due largely to incompetence and the second for its indifference to others' well-being. Previous research has also found almost unanimous public support for mandatory labeling (Einsiedel 2005; Miller 2008). We likewise found essentially no negative responses to labeling in the public groups we studied. This support cuts across age, education, political leaning, and sex.

We should stop for a moment and just consider this result. We have multiple studies that show across demographic and psychographic categories, including across political philosophies, that Americans almost universally support mandatory labeling. This tendency to favor mandatory labeling not only manifests for nanotechnologies but across a wide range of consumer goods. Ross Pifer has noted that we have strong evidence of widespread public support for labeling genetically modified organisms

(GMOs) in the form of ballot initiatives and legislation (Pifer 2014). Consumer demand for such labeling has also produced a rising trend in voluntary labeling of GMOs, as well as policies at grocery outlets like Whole Foods to require such labeling on any products they sell (Wohlers 2013). At the same time, only roughly a third of the population strongly opposes GMOs (Ganiere et al. 2006). Much as we found in this study with nanotechnology, publics want GMO foods labeled, even if they are not interested in heavier regulation. A similar phenomenon emerges in studies of a wide range of agricultural biotechnologies (Juanillo 2001). In some sense, all of these can be explained by the Wiggum-Burns dynamic and its tendency to generate support for labeling over alternative methods of regulation.

Labeling also is entirely consistent with the tendency for publics to have a social perception of regulation, affected by many of the same variables that affect social amplification of risk. For better or for worse, mandatory labeling resonates with many of the core values that drive social perception and amplification. Labeling increases perception of choice, control, and transparency, all of which decrease perception of risk. Relying upon government agencies to regulate more directly, such as by prohibiting products from reaching the market, could actually increase public perception of risk, as the combination of low trust in regulatory bodies would make the decreased control, choice, and transparency of direct regulation generate a social amplification of the perception of risk.

Mandatory labeling also has the unique advantage of being compatible with a wide range of political philosophies. Regardless of one's politics or party affiliations, labeling often draws broad support. From perspectives such as environmentalism, social justice, and consumer protections, labeling makes activism and public campaigns more effective, as well as empowering people over corporations. From free-market, libertarian, and individual freedom perspectives, labeling increases freedom by allowing more choice while also making markets theoretically more efficient by producing greater transparency and more informed purchasing. After all, in order for one to believe that the markets should and can decide whether engineered nanoparticles should be in cosmetics, one needs to give the market (consumers) the choice, which means the awareness of their presence in products.

When applied to consumer products, labeling also cuts through many political philosophies by tapping into NIMBYism (not in my backyard), the desire to keep oneself and one's loved ones out of the sphere of risk while still gaining broader economic or social advantages from development, innovation, or change. Labeling allows one to choose not to be exposed and not to expose loved ones to engineered nanoparticles, while still leaving open the economic and technological advantages of nanotechnology development. While some may find this politically and morally suspect, the tendency exists across the political and socioeconomic spectrum, and will be one more element of public behavior that leads most public groups to support mandatory labeling.

Interestingly, the nearly unanimous support for labeling represents a public recognition of a specific information deficit and a call for additional information to help them be more informed about emerging technologies. Johnson (2009) expresses a common concern among scientists that too much public engagement and public participation in regulation risks science becoming "governed by the ignorance of the mob" (para. 4). Yet, what we actually see is that publics seek information needed to make more informed decisions that matter to them, but find themselves stymied by an inability to acquire key information. That companies refuse to share this information and governments are unwilling or unable to mandate the release of the information only reinforces their perception of the former as untrustworthy and the latter as inept and corrupt. The more that such information is withheld, the more we can expect publics to be guided by the Wiggum-Burns dynamic.

This likewise leads to some important conclusions for future public engagement with nanotechnology specifically and emerging technologies more generally. Because of the previous dominance of the information-deficit model of public engagement with science and public science communication, there has been a dramatic swing of the pendulum in the opposite direction by many researchers toward "upstream" engagement that brings the ideas, knowledge, and values from publics to policymakers and industry leaders (who presumably are upstream of the publics). However, what many of us who conduct such events find (and was the case in all of the public engagement events in this study) is that public groups recognize that when it comes to emerging technologies, they do have an information deficit. All of the public groups want more information to be able to make better decisions, as well as the capacity to make those decisions for themselves (hence the support for labeling).

The solution to the information deficit, however, is not for regulators, government agencies, or industry to simply educate or inform publics (who do not trust government or industry representatives anyway). Likewise, the solution is not to simply have scientists or technology experts educate the public, because these public groups are not seeking education or even scientific literacy. What they want, and what we saw them actively seek out in this study, is empowering information. Public groups consistently wanted information that was actionable, that could make a direct difference in their lives and their choices. That desire was repeated by people in the engagement events in this study in various ways, but especially visible in the one group that decided to do their own research, educate themselves, create their own engagement event, and advocate for their members to take political action.

This auto-engagement group (discussed previously in Chap. 2) is an excellent lens for understanding the kind of information that public groups seek. The group's mission and overall focus was on consumer advocacy and local environmental issues that could affect the health or well-being of community members. After some consideration and research, they decided to focus their event on the use of engineered nanoparticles in cosmetics and sunscreens. None of the research team gave them any guidance or suggestions on topic areas. In fact, after our initial invitation phone call (which lasted less than 15 minutes) they had set about doing their own research and settled on this topic before responding and confirming their participation in the study. They made it clear that they had chosen the topic because it would be meaningful and actually useful to their members.

When they researched the topic and organized their event, they focused on three main themes: potential risks of engineered nanoparticles in cosmetics and sunscreens, how to identify products that might contain engineered nanoparticles, and how to advocate for mandatory labeling or other regulation of engineered nanoparticles in consumer products. Unlike presentations prepared by scientists, the members of the group who made the presentation were comfortable expressing uncertainty about the science and the research, particularly when it came to known and unknown risks. Ironically, that expression of uncertainty is much more consistent with the state of the scientific research at the time and the ways that scientists talked about many of these topics when speaking to other scientists. The presenters in this group also did not feel it necessary to provide a deep background on the definitions, history, or underlying principles of physics relevant to nanotechnologies. Yet, they, nonetheless, were more successful at filling the information deficit that truly matters to public groups.

Based on these findings, I recommend that future public science communications and public engagement with emerging technologies give up the ideas of upstream and downstream, along with the traditional information-deficit models, and work to open a dialogue with any particular public to discover what information they want and is actually useful to them. The role of public engagement should not be to correct what government, industry, or scientists believe is the important information deficit in public groups. That objective is simply not reasonable. We all need our information deficits, we need to focus our attention on the information that matters to us, and we need to be able to filter extraneous information out. This is a fundamental necessity just to function in the world, for the totality of potential information is far too vast and expanding too rapidly for any person to absorb even a meaningful fraction of it.

Yet, we also all have information deficits that we are seeking to fill. Any public engagement with science or emerging technology event will be most effective when it helps to fill those information deficits. However, a researcher or event organizer cannot know what information a particular public desires without first having a deep understanding of that public through a dialogue about what matters to them and what is impeding them from making decisions and choices. Dialogue is the method of discovering a public group's values, as well as finding what information will be meaningful, empowering, and actionable for them. Only then is one ready to begin the process of preparing a public communication or engagement event.

Works Cited

Cobb, M. D., & Macoubrie, J. (2004). Public perceptions about nanotechnology: Risks, benefits and trust. *Journal of Nanoparticle Research, 6*(4), 295–405.

Einsiedel, E. (2005). In the public eye: The early landscape of nanotechnology among Canadian and U.S. publics. *AZoNano: Online Journal of Nanotechnology.* Retrieved from https://www.azonano.com/article.aspx?ArticleID=1468

Ganiere, P., Chern, W. S., & Hahn, D. (2006). A continuum of consumer attitudes toward genetically modified foods in the United States. *Journal of Agricultural and Resource Economics, 31*(1), 129–149.

Hart Research Associates. (2007). *Awareness of and attitudes toward nanotechnology and federal regulatory agencies: A report of findings based on a national survey among adults.* Retrieved from http://www.nanotechproject.org/process/files/5888/hart_nanopoll_2007.pdf

Johnson, D. (2009, January 14). *Nanotechnology and public engagement: Is there a benefit?* [Blog post]. Retrieved from http://spectrum.ieee.org/tech-talk/semiconductors/devices/nanotechnology_and_public_enga

Juanillo, N. K. (2001). The risks and benefits of agricultural biotechnology: Can scientific and public talk meet? *American Behavioral Scientist, 44*(8), 1246–1266.

Macoubrie, J. (2006). Nanotechnology: Public concerns, reasoning, and trust in government. *Public Understanding of Science, 15*(2), 221–241.

Macoubrie, J., & Rejeski, D. (2005, September). Public acceptance of nanotech hinges on trust, confidence. *Small Times,* p. 13.

Miller, G. (2008). Nanotechnology and the public interest: Repeating the mistakes of GM foods? *International Journal of Technology Transfer and Commercialization, 7*(2–3), 274–280.

Pifer, R. H. (2014). Mandatory labeling laws: What do recent state enactments portend for the future of GMOs? *Penn State Law Review, 118,* 789–814.

U.S. Department of Agriculture. (2017). *About the US Department of Agriculture.* Retrieved from https://www.usda.gov/our-agency/about-usda

U.S. Food and Drug Administration. (2013). *Structured approach to benefit-risk assessment in drug regulatory decision-making.* Retrieved from https://www.fda.gov/downloads/forindustry/userfees/prescriptiondruguserfee/ucm329758.pdf

U.S. Food and Drug Administration. (2017). *What we do.* Retrieved from https://www.fda.gov/aboutfda/whatwedo/

Wohlers, A. E. (2013). Labeling of genetically modified food: Closer to reality in the United States? *Politics and the Life Sciences, 32*(1), 73–84.

CHAPTER 5

Lessons for Science Communicators: Assumptions and Assessment

Abstract The results and discussions in Chaps. 1, 2, 3 and 4 detailed the public understanding of nanotechnologies, their discussions, and their responses to various speakers in public engagement events. While these are useful and important insights with implications for future public communication, engagement events, and potential regulation of nanotechnologies, these events and the data we gathered from them also offer guidance for science communicators and public engagement practitioners. This chapter provides that practical guidance in a brief, straightforward way, using both details from the 11 public engagement events in this study and previous research on effective science communication and public engagement with science.

Keywords Science communication • Public understanding of science • Nanotechnology • Public engagement with science

The results and discussions in Chaps. 1, 2, 3 and 4 detailed the public understanding of nanotechnologies, their discussions, and their responses to various speakers in public engagement events. While these are useful and important insights with implications for future public communication, engagement events, and potential regulation of nanotechnologies, these events and the data we gathered from them also offer important guidance for science communicators and public engagement practitioners. This chapter provides that practical guidance in a brief, straightforward

P.J. Gehrke, *Nano-Publics*,
https://doi.org/10.1007/978-3-319-69611-9_5

way, using both details from the 11 public engagement events in this study and previous research on effective science communication and public engagement with science.

Too often, science communicators and public engagement practitioners are thrust into the field with little training or guidance in effective communication practices and the research that supports them. Meanwhile, since at least the 1930s, social scientists in communication have been working to offer actionable insights into the variables that affect audience perception and reaction (cf. Gehrke 2009; Gehrke and Keith 2015). Humanities and liberal arts scholars and educators have been working for much longer on these questions, dating back to the very origin of formal education in arts and science over 2000 years ago. Today, thousands of communication teachers and researchers in colleges and universities in the United States and thousands more in other countries are adding to our knowledge and updating our understanding of how to best communicate. Among these, a significant number deal with scientific, technological, and technical information in a wide variety of contexts. Their findings and advice generally resonate with what we witnessed in this study, though, in a few cases, we found that the context of nanotechnology posed unusual challenges and offered opportunities for refining our thinking.

Models of Effective Communication

When we took these 11 events as a set and analyzed the communication behaviors of the participants, both "expert" and "nonexpert," certain patterns or regularities did emerge, but no model or consistently reliable system of communication could be deduced. This is very much what previous research on science communication and public engagement with science has demonstrated. The reason is that each audience, event, and context is, in some ways, unique, and different events have different goals; thus, there cannot be one right way for all times and places. As Issever and Peach (2010) noted, "there is no standard way of doing things" because every presentation, engagement, or event must be "tailored to the needs of the audience" (p. 1). One of the few things we can say with some certainty across the vast majority of public communication of science and public engagement with science is that what most scientists like and think of as the standard way of communicating does not effectively reach audiences.

Scholars' and researchers' attempts to divide public engagement with science and science communication into models also largely fail to produce practical guidance for communicators and practitioners. When Dominique

Brossard and Bruce Lewenstein (2010) mapped out the dominant models for theorizing about such communication and engagement, they identified four: the information deficit model, the lay expert model, the contextual model, and the public engagement model. When they compared these theoretical models to actual practices of science communication and public engagement with science, they found the events did not fit neatly into discrete categories. In fact, most events of public engagement with science or science communication tend to overlap at least three of these categories. The fact is, as Brossard and Lewenstein conclude, that all these models are complementary, working together in practice rather than competing in the ways they often do in theoretical debates among scholars. Perhaps most interesting is how Brossard and Lewenstein found that, in practice, the contextual model, lay expert model, and public engagement model all fundamentally rely upon or deploy elements of the information deficit model (pp. 32–35).

So where does that leave the public engagement practitioner and science communicator? It leaves us in a position that students and scholars of communication know extraordinarily well, because we have been working with it for our entire careers. There are guidelines for good communication. There are principles that guide effective communication. There are key questions and analytics that should drive people seeking to communicate effectively. However, there are no hard rules, reliable formulae, or even set patterns that are ready-made, off-the-shelf, to be deployed in any given situation. Recipes, stock blueprints, and even previously successful models can all have a place in the preparation of effective communication and public engagement, but they must always be adapted to the communication situation at hand. The ingredients available are different in every communication situation; blueprints must be modified to fit the terrain and space available. Previous models were successful because they uniquely fit the specific details of their communication situation, so they cannot be simply replicated in a new environment or context.

THE COMMUNICATION ECOLOGY

Communication scholar Lloyd Bitzer (1968) proposed a theory he called the "rhetorical situation." Bitzer argued that when we are called to communicate, speak up, or speak out, we are faced with three facets of the situation that should guide our choices: the exigence, the audience, and the constraints. Exigence is the impetus that demands we communicate, which he described as an "imperfection marked by urgency" and is rhetorical to the extent it can be addressed through discourse (pp. 6–7). Audience is, of course, the audience that can take action to resolve that imperfection. Constraints are all the conditions that limit the available means of persuading that audience to effect that resolution. Many scholars have contested Bitzer's theorization or offered updates (e.g., Ede and Lunsford 1984; Vatz 1973), but it continues as a useful tool for training communicators across contexts.

Within the specific context of science communication and public engagement with science (including emerging technologies), Bitzer's theory has a number of gaps and difficulties that make it insufficient. However, with only a few modifications, we can easily produce a more robust model, grounded in more recent research, that offers guidance that is of particular value to public communication and public engagement. Part of the difficulty with Bitzer's position is that it is too wedded to the theories and history of rhetoric, without sufficient concern for broader communication research or contexts. It also depends on metaphors and elements that we understand in much more nuanced ways today than we did 50 years ago. Today, we are better served by thinking about a communication ecology, rather than a rhetorical situation.

Lerner and Gehrke (2018) detailed the principles of ecological thinking and its implications for public engagement with science in a book on organic public engagement, but, in brief, we can understand the shift from the rhetorical situation to the communication ecology as a move from analytic to holistic thinking. That is to say, the communication ecology model moves away from the view of a communication situation as discrete parts, existing and constituted prior to the situation, and then brought into interaction. Instead, communication ecology understands that in every moment of communication the elements are dynamically co-constituting one another, in a state of both solid existence and fluid transition. It also reflects the understanding that because every participant or element in a communication ecology is undergoing change as a part of the communication event, the dynamics and effects can be much more complex than the mechanics of the rhetorical situation model can allow.

Yet, certain analogues to Bitzer's tripartite structure can guide us in preparation for a communication or engagement event and serve as reminders of how we adapt in the moment when the ground shifts under our feet. In lieu of Bitzer's original exigence, audience, and constraints, communication ecology offers purposes, contexts, and externalities. By looking to each of these in order, and how they co-create and mutually affect one another, we lay the foundation for compelling self-presentation, crafting events and messages, delivering them effectively, and working productively in discussions or question-and-answer sessions.

Purposes

Purposes in any given communication ecology constitute a large portion of what we might think of as the participants, audiences, speakers, community members, funding agencies, industry, and other stakeholders. Even within a small local public engagement event or a simple science communication campaign, inevitably, multiple purposes are operative and often competing with each other for priority. Explicit purposes we see for science engagement and communication events include promotion of a technology or product, rousing or quelling public concern or interest in a topic, delivering information (usually with a hope of affecting behavior or opinion), and so forth. But there are usually many additional purposes at play, some reinforcing these, some not. For example, observation of actual public groups interacting with guests or speakers usually shows multiple instances where behavior contains elements of manifesting desires for recognition, affirmation of one's own intelligence or expertise, and similar status seeking. Often, this desire manifests in the behaviors of "expert" speakers whose behavior belies a purpose of demonstrating or validating their expertise.

Other common purposes at play in most communication and engagement events are the desire to be liked, seeking financial gain, bonding with an existing or potential community, performance of a role, and shaming or policing the behavior of others. All of these can have positive and negative value for a communication event, depending on the broader ecology and how they manifest. Two of the most common purposes we observed in the groups in this study were to have a good time and to be part of a group. These are fundamental purposes for most people and manifest by participants in science communication and public engagement with science, regardless of their age, expertise, or education. That such purposes are a major part of the ecology in any science communication or engagement event is both intuitively obvious and very often overlooked. As a result, people planning such events forget that these are key drivers for all the participants.

While it is important to remember that the most common purposes in a science engagement or communication event are to have a good time and be part of a group, this information offers only very vague guidance without being assigned greater specificity. Within different ecologies, what constitutes having a good time or being in the group differs significantly. Even the same people put into different ecologies will find both enjoyment and belonging are achieved differently. Even more, as different communities and events are constituted of different people, everyone's experience of enjoyment and belonging may change. A classroom setting, a community meeting, a nonprofit group, a bowling league, a group of people waiting for a bus, in each ecology enjoyment and belonging will differ.

Likewise, there is no easy set of positives and negatives that we can rank to constitute such enjoyment and belonging, much less other kinds of purposes, that may be more idiosyncratic to a particular ecology. Milton Rokeach's (1973) foundational research did demonstrate that we could find certain commonly shared values that relate to purposes. In fact, Rokeach identified 18 common values, nine of which he labeled as terminal and nine as instrumental. Terminal values are those that are considered good in and of themselves, such as happiness. Terminal values do not need to produce anything else or lead to another purpose in order to be valuable. Achieving a terminal value is, by itself, considered good. Instrumental values, on the other hand, are good because they enable the achievement of another purpose. It may be that an instrumental value helps produce a terminal value, or it might enable another instrumental value that then leads to a terminal value. In any case, the instrumental value is only valuable, according to Rokeach, because it can lead to a terminal value.

We might think of purposes the same way, and one could even use Rokeach's list of values to help analyze the fundamental purposes at work within a certain communication ecology. Of course, a particular communication ecology will also have other purposes at work, more idiosyncratic to that ecology, in addition to Rokeach's fundamental values.

The real insight of Rokeach's work, however, and the part that is most powerful for thinking about purposes and values in a communication ecology, is that these values are not all harmonious to one another. In fact, core values commonly held by the vast majority of people are in fundamental contradiction to one another or placed into contradiction by how we live our lives. So too is the case with the purposes operative in any ecology. Some of these may be classic philosophical tensions, such as freedom and security. Some may be products of lifestyle, culture, and class, such as the tension between work and family or economy and environment. The fact is, though, that every communication ecology is cut through with such tensions and contradictions within its purposes.

These are also not just contradictions between people or between expert and nonexpert participants, or between industry, government, and citizens. While tensions and competing purposes clearly do exist on those levels (as we saw in the previous chapter), they also exist within each of the participants and each of the constituencies or subgroups of a communication ecology. Nearly all of us value both freedom and security, to varying degrees. The same is true for work and family and for economy and environment.

Rokeach saw this fact and noted that the key difference between people is not what they value but how they negotiate and prioritize those values when they compete. He referred to these systems of prioritizing values as value hierarchies and spent much of his career studying and documenting them. Within a communication ecology, participants not only bring these hierarchies, but as we saw in Chaps. 3 and 4, they are constantly negotiating these values and hierarchies with each other, adjusting and shifting them as they encounter both different information and each other. We all live in ecologies of competing purposes and hold competing purposes within ourselves. We choose which to prioritize and which to subordinate not individually but communally and in communication. The Rokeach Value Survey provides a straightforward instrument for studying value hierarchies. It can be used as a survey instrument before a communication or engagement event, or it can be used as an analytic tool when evaluating an ecology's purposes and values, even if you do not have the opportunity to survey the attendees in advance. One can use any public information or literature about the ecology, group, and members to glean values and their hierarchies, especially if those materials were generated from within that ecology by persons who are embedded in the ecology.

Finally, effective science communication and public engagement requires that we respect the purposes and values emergent in the ecology and not just those of experts, scientists, grant-making bodies, or researchers. As Danielle Endres (2010) demonstrated in her study of the public controversy surrounding the Yucca Mountain nuclear waste site, nonscientific values are often equally or even more important than scientific values. In cases of public policy or direct public action, even the "best" course of action based on scientific criteria may not be the best action when all the values come into play. It may well be that, in some cases, the public good requires that scientific values are not the sole drivers of policy or action. In some cases, it may also be that scientific knowledge or expertise overlooks important knowledge and information held in local or nonexpert communities. Brian Wynne's work (1991, 1992, 2003, 2008) is often cited as an example of such a phenomenon. Today, most science communication and public engagement experts agree that, as Massimiano Bucchi (2008) put it, "lay knowledge is not an impoverished or quantitatively inferior version of expert knowledge; it is qualitatively different" (p. 60).

Contexts

The context of any given communication ecology can be understood as physical, administrative, social, and cultural. When I speak of such contexts as a part of a communication ecology, I am not referring to a broader context that encompasses the ecology and other ecologies (which we should classify as externalities). Instead, the contexts that make up this particular ecology, are created by the ecology, and are considered internal parts of the communication ecology. Distinguishing contexts from externalities can require some careful judgment and usually requires some sustained observation of the whole of the communication ecology, but the distinction is not an arbitrary invention of the researcher, practitioner, or communicator. It is set, instead, by the ecology itself, constituted by the purposes and externalities of the ecology as well as co-constituting the other elements of the ecology.

The contexts matter not because they simply set the scene, but because we know that contexts create communication, audiences, behaviors, and attitudes. As discussed in Chap. 1, the research in social psychology and behavioral economics in the past 30 years has demonstrated the impact of context on belief and behavior so consistently and so powerfully that there is no legitimate ground remaining for someone to deny the importance of context in human behavior. Context is not simply the space in which we take action or express belief; it is a major factor in shaping which actions we take, what beliefs we hold, and how we express ourselves.

Analysis of the contexts operative within a given communication ecology may require parsing the contexts into specific subcategories, which may themselves be more or less idiosyncratic to the ecology. At the simplest level, analysis of context should begin by parsing the context into the physical contexts and the social contexts. In some cases, I have also found it useful to separate out cultural contexts or administrative (formal) contexts, when they were significant enough to warrant greater scrutiny than they might receive when folded into social contexts.

Physical contexts include all the physical and spatial elements in which the communication ecology exists and that exist within the communication ecology. That certainly would include a meeting room or building, the layout and dimensions of the room, positioning of the seats and tables, and so forth. But, it also includes things like the colors of the walls, the comfort of the seats, the kind of flooring, the lighting, the sounds and acoustics present in that space, and the movement and position of bodies. It likewise includes the clothing commonly worn, the physical materials present in the room, the presence or absence of food or drink, and every other element that has a physical manifestation. When we analyze and prepare for a communication ecology, we should try to consider as many of these elements as feasible.

This may seem like an onerous level of concern with the physical space, but there are good reasons for trying to account for as much of the physical context as possible. First, there are the obvious and intuitive reasons why physical context can make a significant difference in communication and engagement events. How many people does the space hold? How will they be seated? Will they be able to see the speakers and any visuals being shown? What about acoustics? Will amplification be required? Will participants be able to hear each other if there is an open discussion or question and answer session?

Anyone with even minimal experience planning science communication and engagement events will be familiar with these kinds of questions, but why do we also want to think about things as subtle as wall color, attire, and the presence of food and drink? The reason is that even those elements, which are commonly ignored or thought of as mere background, can have a significant effect on the reception of information, mood of participants, and even the metaphors evoked in the minds of an audience. The evidence for these effects can be found in hundreds of studies over the past few decades, but also is well documented in best-selling popular social-science books such as *Drunk Tank Pink* (Alter 2014) and *Freakanomics* (Levitt and Dubner 2005). It will be impossible to account for every possible variable, but the more one can consider the whole of the physical context in preparing (either as an event organizer or as a speaker), the more likely one will be able to both consciously and unconsciously make effective choices in advance.

There is no better way to account for physical contexts than to actually visit the space beforehand. When I am organizing public engagement events, I make every effort to visit potential sites before making a final selection. When this is not feasible (sometimes because the sites are hundreds or even thousands of miles away), I do the best I can with pictures or video of the space. This is not optimal, but acceptable. When I am an invited speaker, my own practice and the one I recommend to those I coach, is to ask the event organizer to ensure that I can have at least 15 minutes in the room before anyone else arrives in the room. Often, this is done on a previous day if the event is local, earlier in the day if I am traveling to the event, or sometimes just prior to the event before attendees are allowed to enter. Even just 15 minutes gives me time to assess the room a bit and make adjustments to my plan, my presentation, and even possibly a slide or two. Even in cases where I have been sent pictures or video in advance of the event, I have sometimes been surprised to find the room was set up differently from how it was pictured. In one case, the room was completely remodeled and redesigned between when the pictures were sent and the date of the event, requiring some adjustment to my plans for the time allocated. Hotel managers, computer support staff, event organizers, catering services, physical plant operations, and so many others can make changes to the physical context even just hours before arrival. In some cases, there's simply nothing one can do to prepare for all the contingencies, but understanding that the physical context is an important part of the communication ecology and always potentially in flux is a good start.

Social contexts include a host of nonphysical dimensions of a given communication ecology, but, when used by organizers or speakers, it will usually focus on the dynamics of the participants or attendees. At its simplest level, social contexts include basic dimensions of audience analysis, such as demographic data and affiliations. Who are the participants? Do they know each other? Are they members of a shared group or share some common identity? In sum, what can be known about the participants' or attendees' relationships to each other and sense of self-identity.

Beyond this, as the social implies, we ideally would like to have some sense of the kinds of social dynamics that operate among the attendees or participants. Are there cliques or factions? Are they informal or formal? Are they static or more fluid? Is there enmity between them, and, if so, what is its nature? Practically, answers to such questions are very difficult and one can rarely ask an organizer or group leader these kinds of questions directly. Instead, there are essentially three strategies for discovering this kind of data before an engagement or communication event. The first is direct observation over time. At minimum, this means observing the attendees or participants interacting together on multiple occasions before the event. That is possible if one is using the organic public engagement model, even if sometimes impractical. If one is following the principles of organic public engagement, then at least one prior observation should be possible in some cases (assuming time and budget allow).

In addition to direct observation, the second strategy is to interview organizers, contacts, or other familiar with the attendees or participants, using indirect questions. This takes a bit of tact and patience, but one can ask about prior events, responses of attendees or participants, the contact's general sense of the attendees, and so on, to glean some sense of the social dynamics. I have on occasion let the more formal pre-event interview slide into a casual chat about the group and its interesting previous events or important work. In this looser, more conversational mode, it is less odd to ask about whether attendees or participants tend to socialize outside the events (e.g., "It sounds like you have a really great set of regular participants. Do you know if they get together outside the events, too?"). Responses range from pleading ignorance ("I don't know") to revealing that there are some cliques split along particular lines ("Well, a group of the members tend to go out together to a bar after events, and I know some of our other members play golf together") to indicating there is a core in-group and then a large number of one-time or occasional participants ("Some of us who have been with the group for a while and attend regularly like to get together for lunch once a week to talk about the events"). Experience or even training in interviewing for qualitative research can be valuable in this process but is not essential.

Finally, the third common way to glean insight into social context is to have an insider who is willing to work as your informant. While this has been rare in my experience as an organizer of public engagement events, it has been quite common in cases where I have been invited as a participant or speaker. In these cases, one usually can ask quite direct questions. There are two significant risks with such scenarios: First is that the informant may not always be reliable (having either a partial view, a particular bias due to positioning in the group, or both). Second is that too much "insider" information can cause one to appear either too familiar for an outsider or reveal that one of the group members has been revealing information about the group or its members that they are not comfortable sharing.

One element of social context that is usually quite transparent and about which you can easily ask, is the administrative or formal element of the social context. What time do events or meetings start? Is there a regular format or agenda? Who will be speaking in what order for how long?

Is there an expectation or norm to their format? These formalized elements of the social context, which include administration of events, not only help one plan communication or engagement events that fit that social context but also may reveal some qualities about the culture and norms of that communication ecology. Groups with highly formalized structures and agendas are, as a whole, generally quite different to work with than groups that tend to be more free flowing. Some events are organized around a common shared practice, like taking a meal together, and the pacing, scene, and contents of the meal (physical context) significantly affect the communication and the styles of interaction between participants (social context). I have never had anyone display any discomfort when I asked about such formalities, except in admitting they had none (and their discomfort with this fact itself was something to consider as part of the social context). In a few cases, contacts have revealed not only the formal or administrative elements of their social context but then gone on (without prompting) to talk about the differences of opinion within the group about the appropriateness of those formal elements and efforts to change them. I should caution that in such revelations, and especially if a contact, participant, attendee, or organizer asks me to take sides in their internal disputes, I steer far clear of any expression of opinion. I might say something as simple as, "I appreciate you sharing that with me; that's very helpful," affirming their communication without affirming its content. If pressed, I might say, "Well, I obviously don't know the group or have the experience with the participants you and others have. It sounds to me like there are others who are probably in a much better position than I am to have an opinion on what's best."

Finally, analysis of social context should include some dimension of the culture of the group. Here, I refer to common cultural elements that inform group behavior and communication. That may include elements commonly thought of as "culture," such as ethnicity, national origin, or religion, and, in some cases, these elements demand extended consideration. However, in many other cases, the cultural scope of social context might mean the shared stories or myths that inform their thinking. These may well be the origins or foundations that support their purposes and values. In some cases, that means local events that carry significance for the group. In other cases, it quite literally means mythological origin stories. In many cases, one will not fully grasp the whole of the cultural dimensions unless one is embedded in that group for an extended period (and even then perhaps not). For most science communication and engagement events, we cannot hope for deep anthropological knowledge of a group's culture prior to the event. However, we can at least take the time to ensure we, first, have considered that culture might be an important dimension of context and, second, have made a serious effort to account for group culture in our preparation of communication and engagement events.

Externalities

The third and final element of a communication ecology to be considered when preparing science communication and engagement events is the category of externalities. While context encompasses the physical and social elements that make up the given communication ecology, externalities are conditions outside a given communication ecology that may affect that ecology or elements within it. This distinction between externalities and contexts is largely analytic and it can sometimes be difficult to draw the line between them. Still, thinking about externalities as its own category is useful for two reasons: First, it preserves the category of context for considering elements of the ecology, focusing our attention on those elements, and reducing the likelihood that we will overlook important dimensions of the context. Second, it reminds us that every ecology is also a part of a larger ecology and interacts with a number of adjacent ecologies. This both reduces the risk of being surprised by externalities and increases our chance of identifying idiosyncrasies or artificial constraints that might pose challenges in a communication or engagement event.

Externalities may include simple logistics such as the traffic and weather conditions on the day of the event, other events competing for attendees or participants, or limits on the event budget. Externalities may also impose administrative or formal restrictions, such as standards for human subjects research, laws and regulations, or data collection requirements imposed by research methods. Social externalities can also play a role, such as a recent news stories that may affect how participants interact or encounter new information and relationships that attendees or participants may have with people relevant to the event but not a part of that communication ecology. As the saying goes, it's a small world. I have been surprised at how often we find that a science communication or public engagement event reveals affiliations completely unexpected by the researchers and even participants before the event. Twice I have seen nonexpert members of local philanthropic associations identify that they have a direct connection to someone who conducted research of significant relevance to the topic of the event, once being a mentor who featured prominently in an expert participant's presentation. Many years ago, I watched two lawyers representing opposite sides of a legal battle over emerging music-sharing technology bond over their shared love of particular bands and music venues. They likely had both attended the same concert on multiple occasions and never realized it. When they realized this, their relationship to each other shifted dramatically.

Just as with purposes and contexts, externalities tend to be fluid, but often even more so. To what extent these externalities impact an ecology can vary from one ecology to another, depending upon the permeability of that ecology's boundaries and other qualities of a communication ecology. One way to assess such elements of a communication ecology is by analyzing it as a public. Communicators, researchers, and engagement event organizers interested in more details on that approach may find utility in the organic public engagement methodology (cf. Lerner and Gehrke 2018).

Regardless of whether one is planning a robust organic public engagement event, organizing a more artificial model of public engagement, or simply preparing to engage in public communication of science, detailing the purposes, contexts, and externalities provides a foundation for making good communication and engagement choices. Science communicators and engagement event organizers will be well served by looking for opportunities and challenges within the ecology. This is a richer and more reliable model of preparing than conventional audience analysis because it looks at the primary drivers that create audiences and shape their behaviors. As Edna Einsiedel (2008) wrote, audiences and public groups "are products of contexts: the same individual can assume different roles at different times (or the same time)" (p. 175). Einsiedel goes on to document that within different contexts, people behave in significantly different ways (a conclusion well reinforced by cognitive science and behavioral economics research). While Einsiedel is undoubtedly correct, I advise consideration of the broader ecology (including purposes and externalities) as a remedy to the sometime narrow view that can happen when only considering context.

THE SCIENCE COMMUNICATOR'S ROLE AND PERSONA

If you are a participant or speaker at a science communication or engagement event, you likely have been assigned a role as part of the invitation or organization of that event. You might be there as a "public expert" or a "leading researcher" or "industry representative" or "government regulator" or "pubic activist." You may have some control over your role, if the event organizers are open to your input on the matter, but the odds are good you will be constrained by the structure of the event, your positioning by the organizers, and even how you are introduced. You certainly do want to make an effort to have your role defined in a way that not only makes you comfortable, but that seems to fit the ecology as you understand it. How does your role relate to the purposes operative in that ecology? How do the dimensions of the ecology's contexts interact with your role? Are there externalities you should be cautious of or you could capitalize on that are relevant to your role? Every role will have potential positives and negatives.

While you may have minimal control over your assigned role, you have significantly more freedom when it comes to choosing your persona. The persona is the character or qualities you bring to that role, the personality, motivations, values, and dispositions that participants and audience members will recognize as fitting a certain archetype or personality type. As Hans Peter Peters (2008) notes, even if your role is circumscribed to "public expert" you still have a wide variety of options for the persona you will adopt within that role.

Consider just a few of the common personae adopted by scientists in public events: professor, curmudgeon, innovator, revolutionary, whistle-blower, evangelist, apologist, and administrator. The professor tends to adopt a position of teacher and expert, seeking to impart understanding and knowledge to the audience or other attendees (who she or he likely positions as students who need to learn what is going to be taught). The curmudgeon is grumpy, often misanthropic and pessimistic, and may complain about conditions ranging from overregulation to insufficient funding to public ignorance. The innovator is someone doing research that is moving the field forward in important but incremental ways, interested and excited about the progress she or he and colleagues are making. The revolutionary may also be a researcher moving the field forward, but seeks to overturn the conventional wisdom or common assumptions in his or her field. While the innovator will probably express appreciation and a collaborative attitude toward colleagues, the revolutionary tends to diminish the work of colleagues who are not fellow revolutionaries if not be openly hostile to them (as they represent the old way of thinking that must be overthrown). The whistle-blower will depict herself or himself as an insider who will reveal a secret other experts are aware of but covering up. The power of the whistle-blower persona is that the speaker can come across as looking out for others at significant personal or professional risk. Because they position themselves as people who speak truth to power and risk their careers in the process, audiences may associate them with cultural models of similar heroes. For a detailed treatment of the history of this persona, see the analysis of the *parrhesiastes* (the one who speaks the truth) in *The Courage of Truth* (Foucault 2012). The evangelist positions herself or himself less as a researcher and more as a champion for a technology or cause. They may also be innovators, but their persona is more as a public advocate for the technology or cause, preaching its promise and refuting concerns voiced by others. The apologist takes a position similar to the evangelist but focuses on defending an out-of-favor theory or technology, attempting to recuperate its image. The evangelist will be a more common position for new and emerging technologies (like nanotechnology), while the apologist is more frequently adopted in relation to technologies that have a well-known and negative public image (such as coal-fired power plants). Finally, the administrator tends to speak about programs, initiatives, and investments in a technical, neutral, and managerial way. I have seen more than one scientific expert doing innovative research who adopted the persona of an administrator, overviewing the research done in her or his labs, recent and future funding initiatives, people being supervised, and bureaucratic challenges to future work. There may well be times when the administrative persona is appropriate, but in my experience and in all the cases we analyzed in this study, audience and participant response to administrators was at best lukewarm.

Most of us will have a default persona or a few for any given communication ecology. Perhaps I am an innovator in academic publications, a revolutionary at conferences, an administrator in my own university, a curmudgeon when dealing with administrators or funding agencies, a professor with most public audience, and so forth. These defaults may or may not serve us acceptably in many situations, but they are not the ideal and thoughtful way to plan our self-presentation for a science communication or engagement event. When preparing for a science communication or engagement event, we are ideally more deliberate about our choice of persona. Once you have a sense of the ecology and hopefully some details about its operative purposes, contexts, and externalities, imagine yourself in that ecology and think about your default persona. If you have been in similar communication ecologies in the past, use those as a guide for how you would likely behave in this ecology. What would you call that persona? How would you describe it?

Once you have a sense of your default persona, consider its core values and its value hierarchy. It may be helpful to refer to Rokeach's list of instrumental and terminal values. Try to inhabit your default persona for that ecology and consider how you would resolve conflicts between those values within the context of that ecology and your role in it. A serious effort might involve using the Rokeach Value Survey instrument to actually create your persona's value hierarchy and compare it to the results you attained doing the same for the ecology. What you are looking for are congruities and incongruities Looking at the results of comparing your own value hierarchy to the one you generated for the communication ecology, are there problematic incongruities relevant to the topic? Are there potentially opportune congruencies you might capitalize upon? If it seems like this persona is a poor fit for the communication ecology, consider whether you have an alternative persona, perhaps one you adopt in another context or in different kinds of ecologies that you could bring to this event, rather than your default persona. What if you saw this ecology as something more like a gathering of colleagues or a collection of potential issue advocates?

The key is to find an actual congruity between one of your genuine personae and the communication ecology. You need to find a way to address its purposes and tap into the higher-ranked values in that ecology, but the persona must also be genuine. The audience needs to trust you, at least to some extent, and that usually is easiest when you are genuine. You can expect that your assigned role will affect audience trust. The results of this study reinforce the general trend in the research that scientists and scientific institutions (including higher education) are more trusted than most other institutions, such as corporations or government (Peters 2008, p. 140). Yet, your persona can mitigate negative predispositions, enhance positive ones, or potentially even transform you in the eyes of the attendees (Keranen 2010).

At the same time, people rarely if ever get away with pretending to be someone or something they are not. In cases where speakers, panelists, or other participants are announced to even a select few attendees before the event, at least a couple of those people will have done a web search for information about those people. Even if they were not announced in advance, people may well do the same kind of vetting in the middle of the event and, depending on the norms of the ecology, call out disingenuous behavior or use it later after the speaker has departed. In either case, the result is disastrous. We all have multiple personae that we move between as we encounter different ecologies and some mobility in our value hierarchies, depending upon which persona is dominant and the ecology we inhabit. What you want to locate is a persona that is genuine for you and highly conducive to the given communication ecology. That does not mean you need to agree on or share the same content, information, or knowledge, or even agree on what theories or facts are most reliable. What you do need is to have sufficient value congruities and shared purposes so that you can use those to engage the ecology when discussing differences of opinion, examining competing information, or deliberating over a course of action. If you honestly cannot locate high-ranking value congruities, you should expect that there would be little opportunity to actually engage that ecology, communicate effectively within it, or have any kind of productive discussion. In such cases, you are probably best off either declining the invitation to participate or taking a very simple and descriptive approach that will avoid issues of controversy. The stoking of controversy without sufficient shared purposes or values to engage that controversy is only likely to polarize camps and entrench existing opinions.

FRAMING YOUR MESSAGE

Once you have established your ideal authentic persona for the communication ecology, based upon value congruities, you will want to select a frame for your message using that persona and the highest-ranked values of the ecology (ideally ones that are also highly ranked in your persona). Michael Cobb (2005) explains framing as "selecting and highlighting some facets of events or issues over their alternatives and making connections among them with the objective of promoting a particular interpretation or evaluation and a preferred solution" (p. 224). The example he provides is depicting abortion controversies as either "pro-choice" or "pro-life." While, as Cobb notes, issue framing has sometimes been depicted negatively due to associations with elite manipulation of mass opinion, in the context of effective science communication and engagement, we can use issue framing to help connect a given ecology with issues of science and technology. Framing not only can affect people's positive or negative disposition toward a topic, but can also generate or diminish interest.

The simplest and most obvious way to begin thinking about framing an issue for a specific ecology is to consider what subtopics or specific dimensions will be of the most interest in that ecology. For example, in this study, some of the participating public ecologies were composed of philanthropic associations with special interests in areas such as access to clean water, service to the blind, and similar causes. Nanotechnology offers potential applications in many such areas, and highlighting these not only generated greater signs of interest from the participants but also brought responses and behaviors coded as positive. While it may seem intuitively obvious that one should focus on dimensions of a topic most relevant to the given ecology, surprisingly, fewer than half of the invited experts in this study used these as frames for their presentations or comments. Most provided at least some mention of the applications most relevant to the communication ecology's purposes and values, but this was most often submerged in a more general discussion that was less relevant to that ecology.

Beyond this very simple application of issue framing by selecting subtopics, we should also consider how more nuanced issue frames match our chosen persona and the purposes, contexts, and externalities of the ecology. Matthew Nisbett (2010) lists eight common issue frames in public debates about science policy, which offer us a good foundation for planning our own science communication and engagement:

- Social progress (science and technology improving society)
- Economic development/competitiveness (science and technology aiding the economy or reducing risks of foreign competition)
- Morality/ethics (questions of whether certain technologies or scientific advances run afoul of our moral standards)
- Scientific or technical uncertainty (what do we know or can we know)
- Pandora's box/Frankenstein's monster/runaway science (risks of losing control of a new technology or discovery)
- Public accountability/governance (issues of how technology will be channeled to public good or effectively regulated to mitigate risk)
- Middle way/alternate path (offering a new alternative to an existing controversy or entrenched problem)
- Conflict/strategy (how to move from the status quo to a specific end state) (pp. 44–46)

Of course, this is not an exhaustive list of all possible frames, but particularly when developing science communication or engagement, Nisbett (2010) offers an excellent starting point for thinking about how we might set the issue frame for a specific event. Externalities, such as the motives of funding agencies or requirements of research design, may exclude some frames and favor others. Certainly, the purposes and values of a given ecology will be more amenable to some frames. Likewise, the dominant frames need to be compatible with the chosen persona. The Pandora's box frame may be an excellent fit for the whistle-blower persona, but incongruous with the administrator or innovator. On the other hand, the administrator persona may be well suited to a middle-path or conflict/strategy frame.

Conscious consideration of how these elements fit together generates deliberate choices about persona and frame, which then guide the more mundane elements of content and organization in science communication and engagement. Without carefully considering these elements and thinking about how you will weave them together, you risk an event or presentation that is incoherent or, at least, not optimizing its capacity to meet its objectives. Only after completing an analysis of the ecology, choosing a persona, and deciding on the dominant issue frame should one then begin the more mundane task of setting about the development of content, organization, delivery, and visuals that most books on effective science communication emphasize. Given that this ground has been well covered by others, I have no intention of repeating it here except to reinforce that it must be done with care and thought, guided at all times by the criteria established by the ecology, your persona, and the issue frame.

Works Cited

Alter, A. (2014). *Drunk tank pink: And other unexpected forces that shape how we think, feel, and behave.* New York: Penguin Books.

Bitzer, L. F. (1968). The rhetorical situation. *Philosophy & Rhetoric, 1,* 1–14.

Brossard, D., & Lewenstein, B. V. (2010). A critical appraisal of models of public understanding of science: Using practice to inform theory. In L. Kahlor & P. A. Stout (Eds.), *Communicating science: New agendas in communication* (pp. 11–39). New York: Routledge.

Bucchi, M. (2008). Of deficits, deviations and dialogues: Theories of public communication of science. In M. Bucchi & B. Trench (Eds.), *Handbook of public communication of science and technology* (pp. 57–76). New York: Routledge.

Cobb, M. D. (2005). Framing effects on public opinion about nanotechnology. *Science Communication, 27*(2), 221–239.

Ede, L., & Lunsford, A. (1984). Audience addressed/audience invoked: The role of audience in composition theory and pedagogy. *College Composition and Communication, 35*(2), 155–171.

Einsiedel, E. F. (2008). Public participation and dialogue. In M. Bucchi & B. Trench (Eds.), *Handbook of public communication of science and technology* (pp. 173–184). New York: Routledge.

Endres, D. (2010). Expanding notions of scientific argument: A case study of the use of scientific argument by American Indians. In L. Kahlor & P. A. Stout (Eds.), *Communicating science: New agendas in communication* (pp. 187–208). New York: Routledge.

Foucault, M. (2012). *The courage of truth: The government of self and others, II.* New York: Palgrave Macmillan.

Gehrke, P. J., & Keith, W. M. (2015). Introduction: A brief history of the National Communication Association. In P. J. Gehrke & W. M. Keith (Eds.), *A century of communication studies: The unfinished conversation* (pp. 1–25). New York: Routledge.

Gehrke, P. J. (2009). *The ethics and politics of speech: Communication and rhetoric in the twentieth century.* Carbondale: Southern Illinois University Press.

Issever, C., & Peach, K. (2010). *Presenting science: A practical guide to giving a good talk.* Oxford: Oxford University Press.

Keranen, L. (2010). Competing characters in science-based controversy: A framework for analysis. In L. Kahlor & P. A. Stout (Eds.), *Communicating science: New agendas in communication* (pp. 133–160). New York: Routledge.

Lerner, A. S., & Gehrke, P. J. (2018). *Organic public engagement: How ecological thinking transforms public engagement with science.* New York: Palgrave Macmillan.

Levitt, S. D., & Dubner, S. J. (2005). *Freakonomics: A rogue economist explores the hidden side of everything.* New York: Harper Collins.

Nisbett, M. C. (2010). Framing science: A new paradigm in public engagement. In L. Kahlor & P. A. Stout (Eds.), *Communicating science: New agendas in communication* (pp. 40–67). New York: Routledge.

Peters, H. P. (2008). Scientists as public experts. In M. Bucchi & B. Trench (Eds.), *Handbook of public communication of science and technology* (pp. 131–146). New York: Routledge.

Rokeach, M. (1973). *The nature of human values.* New York: The Free Press.

Vatz, R. E. (1973). The myth of the rhetorical situation. *Philosophy & Rhetoric, 6*(3), 154–161.

Wynne, B. (2008). Elephants in the rooms where publics encounter "science"?: A response to Darrin Durant, "Accounting for expertise: Wynne and the autonomy of the lay public". *Public Understanding of Science, 17*(1), 21–33.

Wynne, B. (1991). Knowledges in context. *Science, Technology & Human Values, 16*(1), 111–121.

Wynne, B. (1992). Misunderstood misunderstanding: Social identities and public uptake of science. *Public Understanding of Science, 1*(3), 283–304.

Wynne, B. (2003). Seasick on the third wave? Subverting the hegemony of propositionalism. *Social Studies of Science, 33*(3), 401–417.

MASTER WORKS CITED

Alter, A. (2014). *Drunk tank pink: And other unexpected forces that shape how we think, feel, and behave.* New York: Penguin Books.

Arnett, R. C., Fritz, J. M. H., & Bell, L. M. (2009). *Communication ethics literacy: Dialogue and difference.* Thousand Oaks: SAGE.

Bainbridge, W. S. (2002). Public attitudes toward nanotechnology. *Journal of Nanoparticle Research, 4*(6), 561–570.

Bitzer, L. F. (1968). The rhetorical situation. *Philosophy & Rhetoric, 1*, 1–14.

Bornstein, B. H. (1999). The ecological validity of jury simulations: Is the jury still out? *Law and Human Behavior, 23*(1), 75–91.

Brossard, D., & Lewenstein, B. V. (2010). A critical appraisal of models of public understanding of science: Using practice to inform theory. In L. Kahlor & P. A. Stout (Eds.), *Communicating science: New agendas in communication* (pp. 11–39). New York: Routledge.

Brossard, D., Scheufele, D. A., Kim, E., & Lewenstein, B. V. (2009). Religiosity as a perceptual filter: Examining processes of opinion formation about nanotechnology. *Public Understanding of Science, 18*(5), 546–558.

Buber, M. (1923/2004). *I and thou* (R. G. Smith, Trans.). London: Continuum.

Bucchi, M. (2008). Of deficits, deviations and dialogues: Theories of public communication of science. In M. Bucchi & B. Trench (Eds.), *Handbook of public communication of science and technology* (pp. 57–76). New York: Routledge.

Cobb, M. D. (2005). Framing effects on public opinion about nanotechnology. *Science Communication, 27*(2), 221–239.

Cobb, M. D., & Macoubrie, J. (2004). Public perceptions about nanotechnology: Risks, benefits and trust. *Journal of Nanoparticle Research, 6*(4), 295–405.

© The Author(s) 2018
P.J. Gehrke, *Nano-Publics,*
https://doi.org/10.1007/978-3-319-69611-9

Danzinger, S., Levav, J., & Avnaim-Pesso, L. (2011). Extraneous factors in judicial decisions. *Proceedings of the National Academy of Science, 108*(17), 6889–6892.

Deng, Y., Ediriwickrema, A., Yang, F., Lewis, J., Girardi, M., & Saltzman, W. M. (2015). A sunblock based on bioadhesive nanoparticles. *Nature Materials, 14,* 1278–1285.

Ede, L., & Lunsford, A. (1984). Audience addressed/audience invoked: The role of audience in composition theory and pedagogy. *College Composition and Communication, 35*(2), 155–171.

Einsiedel, E. (2005). In the public eye: The early landscape of nanotechnology among Canadian and U.S. publics. *AZoNano: Online Journal of Nanotechnology.* Retrieved from https://www.azonano.com/article.aspx?ArticleID=1468

Einsiedel, E. F. (2008). Public participation and dialogue. In M. Bucchi & B. Trench (Eds.), *Handbook of public communication of science and technology* (pp. 173–184). New York: Routledge.

Endres, D. (2010). Expanding notions of scientific argument: A case study of the use of scientific argument by American Indians. In L. Kahlor & P. A. Stout (Eds.), *Communicating science: New agendas in communication* (pp. 187–208). New York: Routledge.

Foucault, M. (2012). *The courage of truth: The government of self and others, II.* New York: Palgrave Macmillan.

Ganiere, P., Chern, W. S., & Hahn, D. (2006). A continuum of consumer attitudes toward genetically modified foods in the United States. *Journal of Agricultural and Resource Economics, 31*(1), 129–149.

Gaskell, G., Eyck, T. T., Jackson, J., & Veltri, G. (2005). Imagining nanotechnology: Cultural support for technological innovation in Europe and the United States. *Public Understanding of Science, 14*(1), 81–90.

Gehrke, P. J. (2007). *The ethics and politics of speech: Communication and rhetoric in the twentieth century.* Carbondale: Southern Illinois University Press.

Gehrke, P. J. (2014). Ecological validity and the study of publics: The case for organic public engagement methods. *Public understanding of science, 23*(1), 77–91.

Gehrke, P. J. (2018). Ecological validity. In B. Frey (Ed.), *The SAGE encyclopedia of educational research, measurement, and evaluation.* Thousand Oaks: Sage.

Gehrke, P. J., & Keith, W. M. (2015). Introduction: A brief history of the National Communication Association. In P. J. Gehrke & W. M. Keith (Eds.), *A century of communication studies: The unfinished conversation* (pp. 1–25). New York: Routledge.

Hart Research Associates. (2007). *Awareness of and attitudes toward nanotechnology and federal regulatory agencies: A report of findings based on a national survey among adults.* Retrieved from http://www.nanotechproject.org/process/files/5888/hart_nanopoll_2007.pdf

Honeycutt, J. M. (2011). Dialogue theory and imagined interactions. In J. M. Honeycutt (Ed.), *Imagine that: Studies in imagined interactions* (pp. 193–204). Cresskill: Hampton.

Issever, C., & Peach, K. (2010). *Presenting science: A practical guide to giving a good talk.* Oxford: Oxford University Press.

Johnson, D. (2009, January 14). *Nanotechnology and public engagement: Is there a benefit?* [Blog post]. Retrieved from http://spectrum.ieee.org/tech-talk/semiconductors/devices/nanotechnology_and_public_enga

Juanillo, N. K. (2001). The risks and benefits of agricultural biotechnology: Can scientific and public talk meet? *American Behavioral Scientist, 44*(8), 1246–1266.

Kahneman, D. (2011). *Thinking fast and slow.* New York: Farrar, Straus, & Giroux.

Keranen, L. (2010). Competing characters in science-based controversy: A framework for analysis. In L. Kahlor & P. A. Stout (Eds.), *Communicating science: New agendas in communication* (pp. 133–160). New York: Routledge.

Lave, J. (1997). What's special about experiments as contexts for thinking. In M. Cole, Y. Engstrom, & O. Vasquez (Eds.), *Mind, culture, and activity: Seminal papers from the Laboratory of Comparative Human Cognition* (pp. 57–69). Cambridge: Cambridge University Press.

Lerner, A. S., & Gehrke, P. J. (2018). *Organic public engagement: How ecological thinking transforms public engagement with science.* New York: Palgrave Macmillan.

Levitt, S. D., & Dubner, S. J. (2005). *Freakonomics: A rogue economist explores the hidden side of everything.* New York: Harper Collins.

Li, X., Hu, D., Dang, Y., Chen, H., Larson, C. A., & Chan, J. (2009). Nano Mapper: An Internet knowledge mapping system for nanotechnology development. *Journal of Nanoparticle Research, 11*(3), 529–552.

Locher, J. L., Robinson, C. O., Roth, D. L., Ritchie, C. S., & Burgio, K. L. (2005). The effect of the presence of others on caloric intake in homebound older adults. *Journal of Gerontology, 60*(11), 1475–1478.

Macoubrie, J. (2006). Nanotechnology: Public concerns, reasoning, and trust in government. *Public Understanding of Science, 15*(2), 221–241.

Macoubrie, J., & Rejeski, D. (2005, September). Public acceptance of nanotech hinges on trust, confidence. *Small Times,* p. 13.

Miller, G. (2008). Nanotechnology and the public interest: Repeating the mistakes of GM foods? *International Journal of Technology Transfer and Commercialization, 7*(2–3), 274–280.

Nisbett, M. C. (2010). Framing science: A new paradigm in public engagement. In L. Kahlor & P. A. Stout (Eds.), *Communicating science: New agendas in communication* (pp. 40–67). New York: Routledge.

Nisbett, R. E., & Wilson, T. D. (1977). The halo effect: Evidence for unconscious alteration of judgments. *Journal of Personality and Social Psychology, 35*(4), 250–256.

Peters, H. P. (2008). Scientists as public experts. In M. Bucchi & B. Trench (Eds.), *Handbook of public communication of science and technology* (pp. 131–146). New York: Routledge.

Phillips, A. C., Carroll, D., Hunt, K., & Der, G. (2006). The effects of the spontaneous presence of a spouse/partner and others on cardiovascular reactions to an acute psychological challenge. *Psychophysiology, 43*(6), 633–640.

Pifer, R. H. (2014). Mandatory labeling laws: What do recent state enactments portend for the future of GMOs? *Penn State Law Review, 118,* 789–814.

Robson, K. A. (2002). A review of psychological and cultural effects on seating behavior and their application to foodservice settings. *Journal of Foodservice Business Research, 5*(2), 89–107.

Roco, M. C. (2011). The long view of nanotechnology development: The National Nanotechnology Initiative at 10 years. *Journal of Nanoparticle Research, 13*(2), 427–445.

Rokeach, M. (1973). *The nature of human values.* New York: The Free Press.

Roto, V. (2015). *Ecological UX studies.* Paper presented at CHI'15 workshop ecological perspectives in HCI: Promise, problems, and potential. Seoul, South Korea. Retrieved from http://rizzo.media.unisi.it/EPCHI2015/resources/papers/EcologicalUXstudies.pdf

Scheufele, D. A., & Lewenstein, B. V. (2005). The public and nanotechnology: How citizens make sense of emerging technologies. *Journal of Nanoparticle Research, 7*(6), 659–667.

Shapira, P., & Youtie, J. (2008). Emergence of nanodistricts in the United States: Path dependency or new opportunities? *Economic Development Quarterly, 22*(3), 187–199.

Slovic, P., Fischhoff, B., & Lichtenstein, S. (1986). The psychometric study of risk perception. In V. T. Covello, J. Menkes, & J. Mumpower (Eds.), *Risk evaluation and management* (pp. 3–24). Boston: Springer.

Sundstrom, E. (1975). An experimental study of crowding: Effects of room size, intrusion, and goal blocking on nonverbal behavior, self-disclosure, and self-reported stress. *Journal of Personality and Social Psychology, 32*(4), 645–654.

U.S. Department of Agriculture. (2017). *About the US Department of Agriculture.* Retrieved from https://www.usda.gov/our-agency/about-usda

U.S. Food and Drug Administration. (2013). *Structured approach to benefit-risk assessment in drug regulatory decision-making.* Retrieved from https://www.fda.gov/downloads/forindustry/userfees/prescriptiondruguserfee/ucm329758.pdf

U.S. Food and Drug Administration. (2017). *What we do.* Retrieved from https://www.fda.gov/aboutfda/whatwedo/

van Bommel, W. J. M., & van den Beld, G. J. (2004). Lighting for work: A review of visual and biological effects. *Lighting Research & Technology, 36*(4), 255–266.

Vatz, R. E. (1973). The myth of the rhetorical situation. *Philosophy & Rhetoric, 6*(3), 154–161.

Wohlers, A. E. (2013). Labeling of genetically modified food: Closer to reality in the United States? *Politics and the Life Sciences, 32*(1), 73–84.

Wynne, B. (1991). Knowledges in context. *Science, Technology & Human Values, 16*(1), 111–121.

Wynne, B. (1992). Misunderstood misunderstanding: Social identities and public uptake of science. *Public Understanding of Science, 1*(3), 283–304.

Wynne, B. (2003). Seasick on the third wave? Subverting the hegemony of propositionalism. *Social Studies of Science, 33*(3), 401–417.

Wynne, B. (2008). Elephants in the rooms where publics encounter "science"?: A response to Darrin Durant, "Accounting for expertise: Wynne and the autonomy of the lay public". *Public Understanding of Science, 17*(1), 21–33.

Zhang, Y. (2008). The effects of perceived fairness and communication on honesty and collusion in a multi-agent setting. *The Accounting Review, 83*(4), 1125–1146.

INDEX

A
Auto-engagement, 28, 61

B
Behavioral economics, 2, 4, 7
Bitzer, Lloyd, *see* Rhetorical
    situation
Buber, Martin, 32–33

C
Coding, 14, 15, 41
Communication ecologies, 67–84
    context of, 72–77
    externalities, 77–78
    purposes in, 69–72

D
Department of Agriculture, 55
Dialogue, 32, 34, 61
Disposition score, 41–44

E
Ecological validity, 5–8, 10
    versus external validity, 6
Experts, 30–37, 47, 66, 70, 78–82

F
Food and Drug Administration, 55
Framing, 82–84

G
GMOs, 59
Gray goo, 35

I
Information deficit model, 60–62, 67
Instrumental values, 70

M
Monologue, 31, 32, 34, 36

© The Author(s) 2018
P.J. Gehrke, *Nano-Publics*,
https://doi.org/10.1007/978-3-319-69611-9

N
Nanomaterials, *see* Nanotechnology
Nanoparticles, *see* Nanotechnology
Nanotechnology, 1–4, 8–19, 21–30,
    33–35, 39–49, 51–62, 66, 82
  and cosmetics/sunscreens, 3,
    44–46, 58
  definition of, 3, 23
  government regulation of, 51
  and industry/private sector, 55–57
  labeling of, 58–60
  public awareness of, 3, 4, 15, 22, 24
  public understanding of, 22–28
  risk perception of, 18, 19, 46–49,
    52, 59, 61
  sentiments towards, 39, 54
    (*see also* Disposition score)
NIMBYism, 59
NSF grants, 8
NVivo, 15

O
Organic public engagement, 8, 68,
    75, 78

P
Persona, 79–82
Public engagement, 7–9, 15, 22,
    29–32, 36, 40, 46, 47, 52, 55,
    60–62, 66–70, 77–84
  See also Organic public engagement

R
Rationality, 5
Rhetorical situation, 67

S
Sampling, 9–11
  deliberate, 8, 11
  random, 10
Science communication, *see* Public
    engagement
Self-selection bias, 10
Study design, 5–6
Survey research, 4, 13, 22–27, 29
Synergy, 36

T
Technical dialogue, 31, 32, 34, 35
Terminal values, 70

U
Upstream engagement, 29, 31, 32, 36,
    46, 60, 61

W
Wiggum-Burns dynamic, 57–58